SECURITY ELECTRONICS

by

John E. Cunningham

Howard W. Sams & Co., Inc.
4300 WEST 62ND ST. INDIANAPOLIS, INDIANA 46268 USA

Indianapolis, Indiana 46268

SECOND EDITION
THIRD PRINTING—1980

International Standard Book Number: 0-672-21419-9
Library of Congress Catalog Card Number: 76-62718

Printed in the United States of America

Preface

Burglary is big business today. Police and insurance-company records show that burglars, holdup men, and vandals are costing the American public well over a billion dollars a year. This does not include the untold loss in human injury and death that cannot be expressed in dollars. To curb this rising toll, Americans are spending in excess of 250 million dollars a year for electronic security equipment. The bulk of this equipment consists of various types of electronic intrusion alarms. These range from simple home and automobile burglar alarms to complex systems designed to protect an entire industrial plant or a school system.

This book describes the principles of operation of various electronic security devices and systems. Their special features are covered in sufficient detail to enable the electronics man to apply his knowledge and experience to the security field.

Since the first edition of this book appeared in 1970, crime certainly hasn't diminished, but many advances have been made in the electronic systems that are used to combat crime. In 1970 many security systems used vacuum tubes; transistors were just finding their way into new systems. Now the vacuum tube has practically disappeared from the scene, and most modern systems use integrated circuits. To reflect these advances, it has been necessary to revise most of the chapters in this book.

Chapter 1 is a general description of electronic intrusion alarms. Chapters 2 through 7 describe the principles of different types of intrusion detection devices. Object detectors, the subject of Chapter 8, were used on only a limited basis when the first edition was published. Since then, they have become widely used to combat shoplifting, and metal detectors are familiar to all who travel by air. These changes have required a complete revision.

Chapter 9 describes the different alarm and signaling devices, and Chapter 10 covers other accessories such as power supplies and access switches that are common to all systems. Chapter 11 gives examples of practical installations. Chapter 12 covers bugging, debugging, and speech scrambling systems.

A new Chapter 13 on automobile protection has been added to this edition. Completely new systems have appeared that are used to identify personnel automatically and to help in verifying the truthfulness of statements. These are covered in new Chapters 14 and 15.

Finally, the computer has had a profound impact on security in two ways. First, computers, particularly microprocessors, have become integral parts of security systems. Second, computers have become the victims of crime. The first aspect of computers in security systems is covered in a new Chapter 16. Protecting the computer from industrial espionage is covered in Chapter 12.

A new Appendix has been added covering terminology used in the security field as published by the Law Enforcement Assistance Administration of the Federal Government.

JOHN E. CUNNINGHAM

Acknowledgments

The author wishes to acknowledge the helpful cooperation of his many coworkers in the security-electronics field, particularly

Howard Reed, American District Telegraph
Jerald Avery, Alarmtronics Engineering, Inc.
C. N. Williams, American Electronics, Inc.
James L. Barnes, George L. Barnes and Associates
John Weinrich, Continental Instruments Corp.
M. M. Wilson, Euphonics Marketing
Frank A. Arneson, GTE Sylvania, Inc.
Tom Probst, Honeywell, Inc.
Lawrence J. Dolan, Identimation Corp.
M. M. Schwartz, Infinetics, Inc.
W. E. Keefe, Johnson Service Co.
Stanley F. Cooper, Walter Kidde and Co., Inc.
Bruce Loughry, Jr., KMS Watermark Systems
M. N. Barwich, Minilert Co.
Archie J. Baley, Notifier Co.
Richard J. O'Conner, Pinkerton Electro-Security Co.
William Ayala, Radar Detection Systems, Inc.
R. W. Lorenz, Red Owl Stores
N. Thomas Berry, Squires Sanders, Inc.
Steve Eldard, SRS Systems
Ron Woodruff, Towmotor Corp.

Contents

CHAPTER 12

CHAPTER 13

CHAPTER 14

CHAPTER 15

CHAPTER 16

APPENDIX A

CHAPTER 1

Electronic Intrusion Alarms

In the past, the word "electronics" had an almost magic connotation. The mere fact that a store or other business was protected by an "electronic" system was enough to discourage all but the most intrepid burglar. Many small businessmen took advantage of this fact and, without bothering to invest in the actual system, merely displayed signs or window decals that indicated that an electronic intrusion alarm was in use. Dummy systems such as imitation closed-circuit tv cameras, complete with lenses and flashing lights, are still to be found in some establishments. These devices are just empty boxes. They might frighten away an amateur, but to the skilled burglar of today they are nothing but an invitation to break in with impunity.

There is a tendency for the average, law-abiding citizen to think of a burglar, holdup man, or vandal as a rather stupid person. This simply is not true. The burglar today is a skilled craftsman, equipped with up-to-the-minute tools for his trade. His stock in trade often includes a detailed knowledge of the principles of intrusion alarms and techniques that can be used to foil them. This has led to a striking difference between electronic intrusion alarms and other types of electronic equipment. Whereas most commercial electronic equipment must work as reliably as possible with the help of skilled electronic technicians, an intrusion alarm must operate reliably, even with a technician doing his best to keep it from working.

TYPES OF ALARM SYSTEMS

Many different types of intrusion alarms, using different operating principles, are available. Each type has its own advantages and limitations. The selection of an alarm for a particular installation is based on the following considerations:

1. If possible, burglaries, holdups, and acts of vandalism should be prevented from happening. The presence of an effective alarm system is definitely a deterrent to would-be burglars and vandals. No one wants to break into a home or establishment where all previous intruders have been caught and sent to jail.
2. The presence of an intruder must be detected as early as possible. There is a saying in the security field to the effect that a burglar can open *any* safe or vault if he is given enough time.
3. The alarm must bring a quick response. An alarm is useless unless action is taken.

Intrusion alarm systems are usually classified in one of three general categories:

1. Proprietary alarm systems
2. Central station alarm systems
3. Local alarm systems

The principal features of each type of system are described in the following paragraphs.

Proprietary Alarm Systems

A proprietary system is one in which the presence of an intruder in any protected area of a facility is indicated at some central location, such as the headquarters of the security guard. This type of installation is widely used in industries and in institutions such as schools and manufacturing plants that have their own security police forces. The facility may include many different buildings that require different degrees of protection. Each area is equipped with the type of alarm system best suited to the application. The signal outputs from the individual areas are then connected by wires to a control panel at the central location. A typical central station console is shown in Fig. 1-1.

With a properly designed installation of this type, one man can monitor the security status of an entire industrial plant. When an intruder enters any of the protected areas, a signal is flashed to the central console. One or more security guards can be immediately dispatched to the scene.

Proprietary alarm systems are often used for much more than intrusion alarms. The central console is arranged to provide an indication (and often a printed record) of such things as the operation of fire sensors and guard patrol stations. As a guard patrols a facility, a record of the time that he checks in at each patrol station along his route is recorded. If the guard encounters trouble and takes more than the scheduled amount of time in patrolling, an alarm will be sounded so that another guard may be dispatched to help him. This type of proprietary alarm system is often called a Supervisory Alarm System.

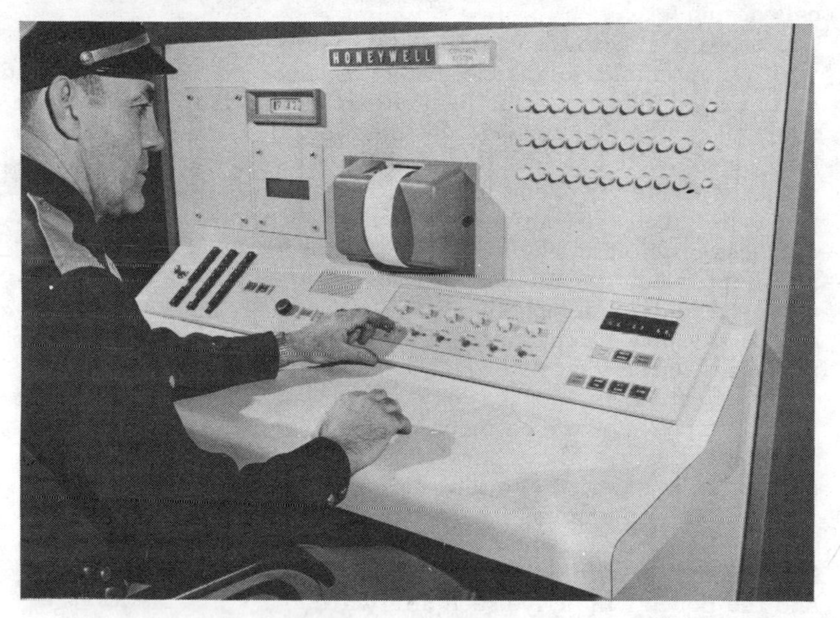

Fig. 1-1. A typical centralized-protection control console.

In many plants the supervisory system will maintain a printed record of each time that a gate is opened or closed. Thus, any unauthorized entry can be pinned down.

Central Station Alarm Systems

Facilities that have limited or no security forces of their own may use remote alarm systems. In this system, each of the areas to be protected has its own intrusion detector. The outputs of the systems are wired to the headquarters of a private security company.

The connection to the remote location may be provided by a radio link, leased wires, or regular telephone lines.

When an alarm is received at the security company, a guard is dispatched to the scene; usually the police are called as well. Many central station alarm systems have audio monitors so that when an alarm is received, the guard at headquarters can actually listen to what is happening in the protected premises. In this way, an unnecessary call to the police department can be avoided if there is no aural evidence of an intrusion. For example, the alarm may be triggered by a thunderstorm. The person monitoring the system can hear the thunder. When there is no evidence of footsteps or breaking into safes or cabinets, the police need not be called.

Closely related to the central station alarm system is the system with an automatic telephone dialer. In such a system, the alarm is connected to a device that will automatically dial a number and transmit a recorded message. In some localities automatic dialers may be connected so that they dial the police department. Sometimes, separate telephone lines are run to the police department just to receive calls from automatic dialers. In other localities, ordinances prohibit arranging dialers so that they dial the police department telephone number. In these locations, the dialer can be arranged to dial a telephone answering service that will relay the call to the police department. Fig. 1-2 shows an automatic telephone-dialing service.

The effectiveness of this type of system depends on how long it takes the police or the security service to respond and get to the scene of the crime. Many burglaries netting large sums have been performed in a period of a few minutes.

Remote alarm systems are often used as a backup to the proprietary systems previously described. With this arrangement, the local security guards will respond immediately to any intrusion, and the police will arrive soon afterward.

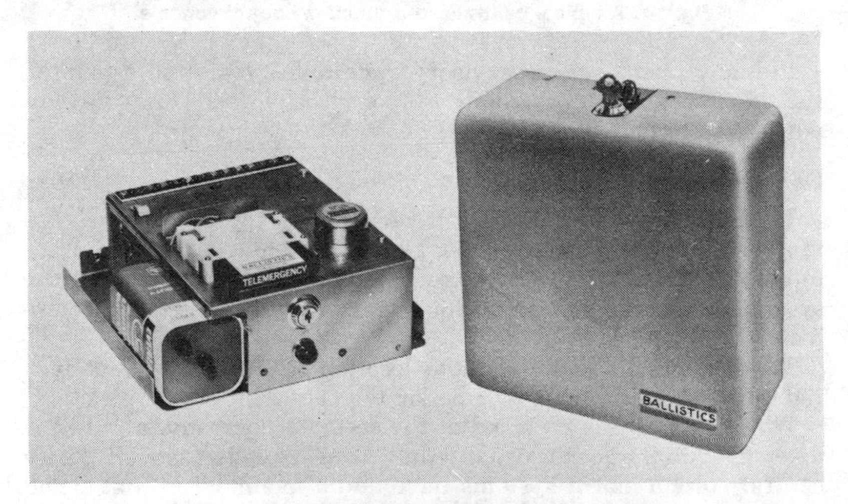

Fig. 1-2. An automatic telephone-dialing device.

Local Alarm Systems

Local alarm systems, as the name implies, sound a bell or siren on the premises whenever an intruder trips the alarm. These systems are usually employed where the other systems are not practical. For example, a local alarm would be used in a remotely located facility where the time required for the police or representatives of a security service to arrive would be so long that they would invariably reach the scene after the burglar had gone.

One disadvantage of the local alarm is that the burglar knows when his presence has been detected. He knows he can stay only one or two minutes before he must leave. Many local alarms bring a response only after the intruder has gone.

In spite of this limitation, the local alarm has its applications. For example, a homeowner may not be willing to pay for a large system that will bring the police, but may feel that he can adequately defend his family and possessions if he is merely alerted to the presence of an intruder. A well-designed local alarm will detect an intruder when he starts to enter, and will alert the resident. In many neighborhoods, people have a great enough sense of responsibility that they will call the police if they hear a burglar alarm in the area.

One factor that should not be ignored when considering the local alarm is its ability to produce a strong psychological effect. For example, an intrusion alarm arranged to activate an ear-splitting siren will unnerve most intruders. In addition to the disturbance, the siren will make it impossible for the intruder to hear the approach of a police cruiser. This along might force a burglar to leave before he has finished his planned activities. The psychological effect can be enhanced considerably by adding flashing, blinding lights to the audible alarm.

TYPES OF PROTECTION

Most of the chapters of this book describe the detailed principles of operation of different types of intrusion alarms. The type of alarm that is best suited to a particular installation depends to a great extent on the exact type of protection that is required and the extent of protection that is economically feasible. The action taken by a prospective intruder usually depends on his expected reward. If the prize is great enough, he will resort to any plan that seems to have a chance of success, no matter how elaborate the plan may be. He will make every attempt to frustrate any detection system and will take great risks. On the other hand, if the maximum possible return from breaking into a home or business is probably small, he will not be willing to take such great risks, and a comparatively simple system might be an adequate deterrent.

Three general categories of protection are available:

1. Perimeter, or point-of-entry, protection
2. Specific-area protection
3. Spot protection

One or more types should be considered for every installation.

Point-of-Entry Protection

Where practical, it is usually advisable to detect an intruder as early as possible. This is usually accomplished by installing detectors on doors, windows, gates, and fences. The system is designed to initiate an alarm as early as possible, before the intruder has a chance to accomplish anything. Stores and other places of business that are closed during the night usually require this type of protection.

The principal limitation of this arrangement is that it is rarely practical to provide complete protection. Even if all the doors, windows, and even walls of an area are protected, it is still possible, and not at all uncommon, for burglars to enter an area by cutting through the floor or ceiling.

There are also facilities where perimeter protection is not practical. It might not be desirable, for example, for an alarm to sound every time an intruder passes through a freight yard. Similarly, a facility that is open for business twenty-four hours a day has no use for perimeter protection.

Another limitation of perimeter protection is that it is useless against the "stay-behind." The "stay-behind" is a burglar who enters a place of business during normal business hours. He then finds a hiding place and remains hidden until after the place is closed and all employees have gone. When he feels that it is safe, he comes out of hiding and helps himself to the merchandise. Then he breaks out of the store, tripping the perimeter alarm on his way out. By the time the police or security guard arrives, he is well on his way.

Specific-Area Protection

Specific-area protection uses systems that detect the presence or movement of an intruder in an area. This type of protection is an excellent addition to a perimeter system. It will pick up the "stay-behind" quickly as soon as he decides that all is clear and starts to move around.

Specific-area protection systems are frequently used where a particularly sensitive area needs extra protection and where perimeter systems are not practical. For example, part of a factory may operate twenty-four hours a day while other parts are only open during the daytime. A perimeter system would be useless because it would be tripped whenever a legitimate worker entered the plant. In such a case, the areas that are not normally occupied at night can be equipped with a specific-area protection system. Thus, if anyone attempts to enter a closed area such as an office or stockroom, an alarm will be actuated.

Spot Protection

A spot-protection system is usually associated with one or more specific objects, such as a safe or a jewelry case. It trips an alarm whenever anyone touches, or in some cases even comes near, the protected object. This type of system is used to back up other systems to provide maximum protection for highly sensitive objects. File cabinets containing secret data, safes, and cases containing valuables are often protected in this manner.

Spot-protection systems are often used as annunciator systems during normal business hours. For example, the proprietor of a store may be happy for customers to browse through his store, but would want to know if they attempted to open a shadowcase containing valuables. A system that is connected to a regular system during the night can be connected to a small bell or buzzer during the day. In this way, there would be a warning whenever anyone tried to open the protected case. Similar systems can be used to protect file cabinets containing government or business secrets.

Intrusion Detection

Every electronic intrusion alarm must have some means of detecting the presence of a human being in the protected area. The part of the system that accomplishes this is called a detector or sensor.

The ideal intrusion detector responds only to the presence of a human being and not to the presence of animals such as dogs, cats, or even mice or rats. It should not respond to any normal changes in ambient conditions such as temperature, humidity, wind, rain, sound level, or vibration. Unfortunately, most devices that will respond to the presence of a human being may also respond to one or another of these extraneous influences. The art of selecting and installing an effective intrusion alarm is to tailor the system to the application so that it will not respond to anything other than an intrusion and, at the same time, will respond to every intrusion.

There are several properties of a human intruder that can be used as the basis of an intrusion alarm system. Probably the most obvious is that an intruder must remove barriers before he can enter the premises. He must open a door or window, or cut a hole in the walls, floor, or roof. Switches can be arranged so that they will be actuated by any attempt at entry.

Another property of a human being that can be used as the basis of an intrusion alarm is the fact that a human is opaque to light and infrared emission. A photoelectric system can be arranged to detect the passage of an intruder.

A less obvious, but quite effective, alarm system depends for its operation on the fact that a human being emits infrared energy due to normal body heat. Such a system has infrared sensors that will detect the heat radiated by a human being.

Yet another property of a human being that is taken advantage of in intrusion alarms is the fact that in many facilities an intruder cannot steal anything or do any damage without making a lot of noise. An audio alarm will detect any sounds that are louder than the normal sounds in the protected facility.

Two very popular alarm systems operate on the principle that motion of a human being will disturb either an ultrasonic or an electromagnetic field. Ultrasonic alarms and microwave intrusion detectors operate on this principle.

FALSE ALARMS

The requirement for extreme reliability in the face of attempts to sabotage the system has led to the design and installation of systems that are "fail safe." In such systems, the alarm will sound whenever the power or any of the components in the system fails. On the surface, this appears to be an advantage, but if the number of false alarms is high, it can actually be a disadvantage because it will destroy confidence in the system.

In some instances, store managers have been called out of bed at night so often by false alarms that they no longer use the alarm system. Nobody has confidence in a system that is often wrong. It is like the story of the boy who cried "wolf" too often.

There are cases on record of burglars using false alarms to foil an otherwise foolproof alarm system. In this scheme, the burglar regularly trips the alarm and leaves immediately. The police or guards arrive, only to find no trouble. After the burglar has done this long enough to completely undermine everyone's confidence in the system, he breaks in and helps himself. For this reason, well-designed systems differentiate as much as possible between an actual intrusion and a component failure.

CHAPTER 2

Electromechanical Detectors

The simplest type of electronic intrusion alarm consists of a closed circuit around the area to be protected. An intruder entering the area will, at least in theory, break the circuit and set off the alarm.

A typical circuit for an alarm of this type is shown in Fig. 2-1. Normally, the 1.5-volt battery keeps transistor Q1 cut off so that there is negligible current in its collector circuit. Since the transistor is cut off, there is only a very small current, usually less than a milliampere, in the protective circuit. When the protective circuit is broken, the reverse bias from the 1.5-volt battery is removed and the transistor is biased in the forward direction through resistor R1. This causes the transistor to conduct and the voltage at its collector to drop rapidly. Since the collector is connected to the negative side of the supply, its voltage actually becomes more positive. This positive-going signal is coupled through capacitor C1 to the gate of the SCR, causing it to fire. The resulting voltage drop across resistor R4 actuates the alarm. Once the SCR has fired, it will continue to conduct until the circuit is reset by momentarily interrupting the power supply.

This circuit cannot be foiled merely by placing a jumper across the protective circuit. Suppose, for example, that an intruder attempted to place a jumper between points A and B in the circuit, hoping that he could then break the protective circuit without tripping the alarm. If this were done, current would flow through resistor R1 and diode D1 to ground. Diode D1 is a silicon diode that has a forward voltage drop of about 0.7 V. Transistor Q1, on the other hand, is a germanium transistor that has a base-to-emitter voltage of only about 0.2 V. Thus, the voltage drop across

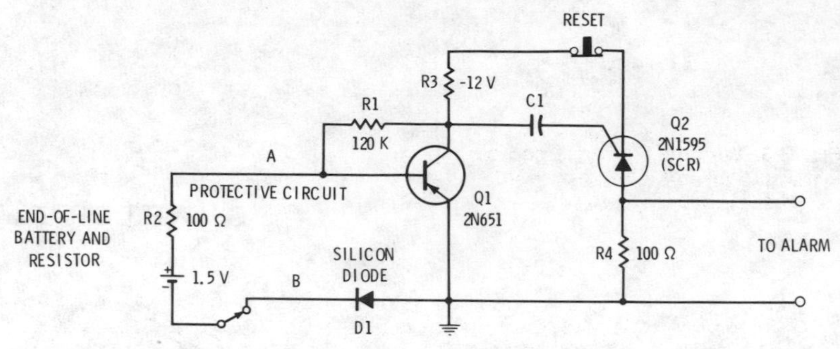

Fig. 2-1. Typical electromechanical intrusion alarm.

diode D1 would be great enough to turn on transistor Q1 and trip the alarm.

TYPICAL DETECTORS

The electromechanical detector is the oldest type of electronic intrusion detector, and its effectiveness depends on the type of detector that is used and how well it is applied to the particular problem.

The electromechanical detector most commonly seen is a metallic foil or tape applied to windows and doors in such a way that an intruder will break the foil when attempting an entry. The effectiveness of foil depends on how well it is installed and maintained. Unfortunately, in many instances, the foil is installed as shown in Fig. 2-2. Here the points where the foil enters and leaves the window are obvious. A skilled burglar would instantly realize that if he managed to connect a jumper between points A and B, he could then break the window without disturbing the circuit. A skilled operator can usually cut a hole in the window large

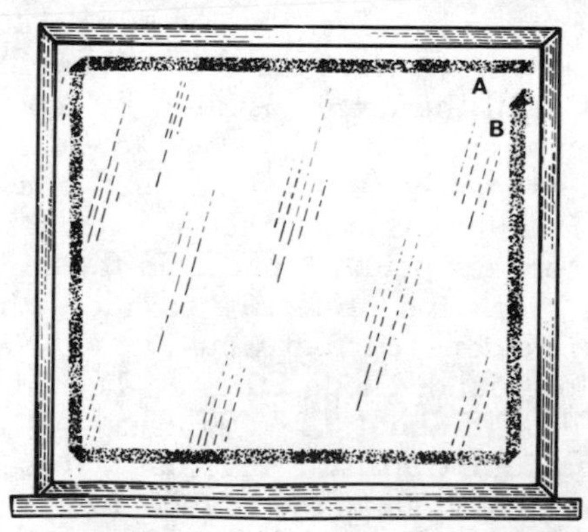

Fig. 2-2. Poor window foil arrangement.

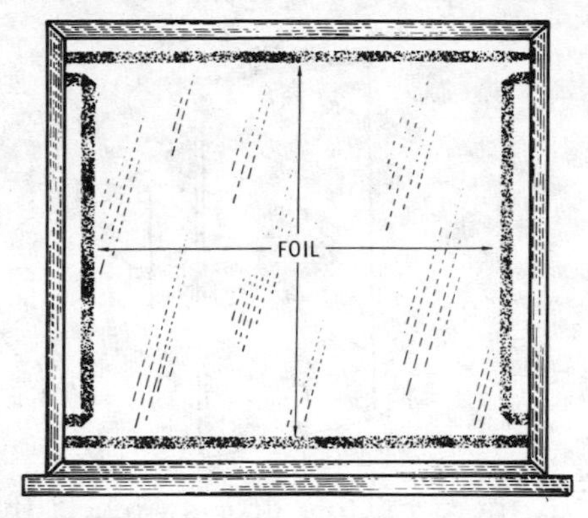

Fig. 2-3. Window foil arranged so that the circuit is not apparent to the intruder.

enough to permit installing the jumper. It takes only a little more time to install the foil as shown in Fig. 2-3. Here, the actual configuration of the circuit is not at all obvious. Installing a jumper between two sections of the foil might actually trigger the alarm rather than frustrate it.

The most common problem with foil or tape is that it is not properly maintained. The foil is subject to wear and abrasion from normal window washing, and after a while it will become ragged and break. Usually the break occurs at a most inconvenient hour and causes a false alarm.

For foil to be effective, it must be inspected frequently and any section that shows signs of wear must be replaced.

Connections to foil should be made through blocks designed for the purpose. If connections are made by soldering lead wires directly to the foil or by soldering the end of the foil to a wire, the connection will be unreliable and a potential source of false alarms.

Several very reliable door and window switches are available for use with electromechanical intrusion alarms. The Law Enforcement Assistance Administration (LEAA) has published standards governing such switches. Switches that meet these standards will prove to be reliable and not subject to changes in ambient conditions.

The plunger-type door switch, shown in Fig. 2-4, is commonly used to initiate an alarm when a door is opened. To be effective, the switch must be installed so that it cannot be seen or tampered with. This is usually accomplished by mounting the switch on the jamb side of the door, as shown in Fig. 2-5. Even with this arrangement it may be possible for an intruder who is aware of the location of the switch to slip a thin piece of steel between the door and the jamb in such a way as to keep the plunger depressed when the door is opened.

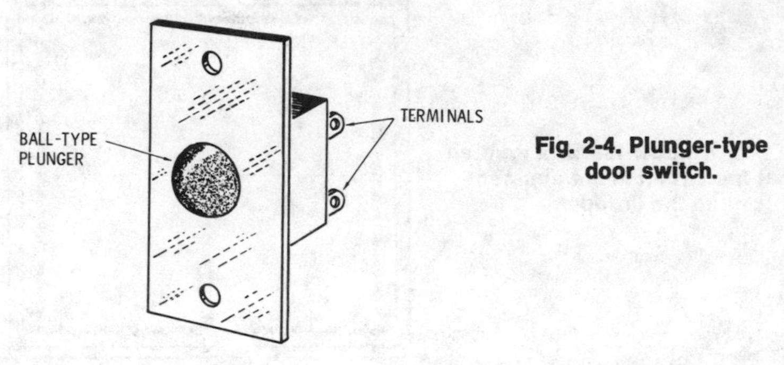

Fig. 2-4. Plunger-type door switch.

A more reliable door switch is the magnetic switch shown in Fig. 2-6. Here a reed switch operated by a magnet is mounted on the door casing, and the operating magnet is mounted on the door. The usual arrangement is for the magnet to keep the switch closed

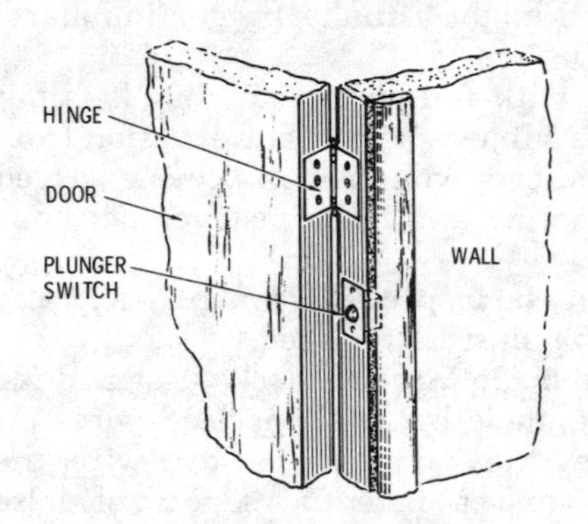

Fig. 2-5. Plunger switch mounted behind a door.

while the door is closed. Then when the door is opened, the protective circuit will be opened and the alarm will be triggered.

Many different types of magnetic door and window switches are available for various applications. The number of places where switches can be installed to detect tampering or intrusion is lim-

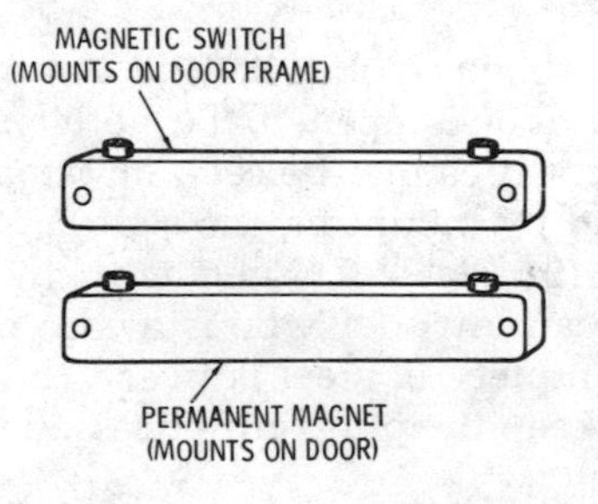

Fig. 2-6. Magnetic door or window switch.

ited only by one's imagination. Switches can even be installed in cash registers so that when money is removed, the switch will be actuated.

Another frequently used switch is a pressure-sensitive floor mat that will open a switch if it is stepped on. This arrangement is often used to provide spot protection for sensitive objects such as file cabinets, safes, or jewelry cases. If an intruder steps on the mat, he will set off the alarm.

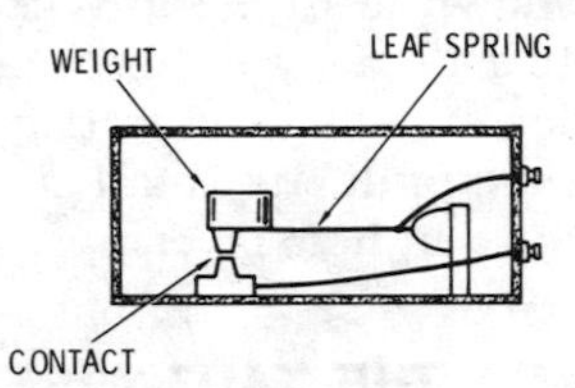

Fig. 2-7. Vibration-actuated switch.

TILT AND VIBRATION SWITCHES

Switches are available that will open or close whenever they are tilted or vibrated. The vibration switch has a mass or weight suspended on a spring as shown in Fig. 2-7. When the switch is at rest, the switch will be closed. When the switch is vibrated, the mass will move and momentarily open the switch and trigger the alarm. Vibration switches are frequently used to protect automobiles. The switch will open if the car is started or if it is jacked up so that a tire can be stolen.

Mercury switches are often used to initiate an alarm when they are tilted. Thus, a mercury switch located inside a safe or cabinet will not be actuated if a door or drawer is opened in normal use, but will initiate an alarm if an attempt is made to carry it off.

Fig. 2-8 shows a self-contained intrusion alarm that can easily be adapted to many different applications. The unit shown contains its own amplifier and alarm oscillator as well as a small

Fig. 2-8. Self-contained electromechanical alarm.

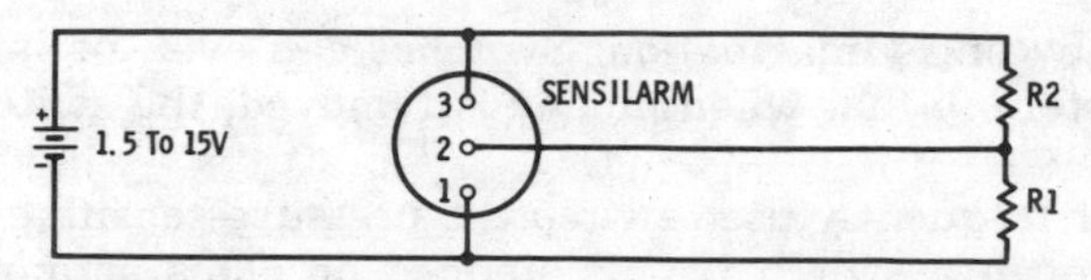

Fig. 2-9. Circuit of the alarm pictured in Fig. 2-8.

speaker and triggering circuits. The connection diagram for the unit is shown in Fig. 2-9. With the connections shown, the alarm can be triggered by either increasing R2 or decreasing R1. Thus, the device may be used to trigger either by closing a circuit or by opening a circuit. Its small size makes this unit particularly suitable for use in homes, camps, and offices.

THE TAUT-WIRE INTRUSION ALARM

Another form of electromechanical intrusion alarm is shown in Fig. 2-10. This system consists simply of a taut wire strung around the area to be protected. The wire, which may be so small as to be nearly invisible, is connected mechanically to a snap-action switch in such a way that the switch will be thrown when the tension is either increased or decreased. Thus, the alarm will go off if an intruder brushes against the wire, or if he cuts it.

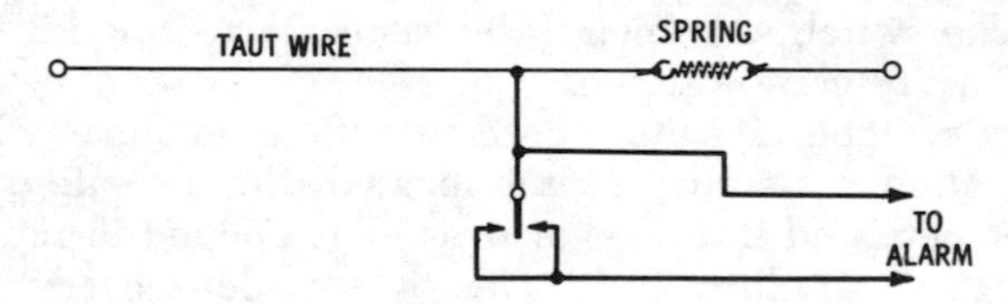

Fig. 2-10. Taut-wire alarm system.

In the simple form shown, the taut-wire system would be subject to many false alarms due to expansion and contraction of the wire. Commercial units are available that automatically compensate for temperature changes.

LIMITATIONS

The major limitation of the electromechanical intrusion detector is that it is usually not practical to protect all possible avenues of approach to the protected area. Even if all of the doors and windows are protected, it is still possible for an intruder to enter through the walls, roof, or floor. He will attempt this if the expected reward is great enough. For this reason, the electromechanical system is rarely used when maximum security is required.

Another limitation of many systems of this type is that they are poorly installed, without imagination. The sensors are clearly visible, and with a little ingenuity they can be frustrated. Of course, this is a limitation of the particular installation and not of the principle itself.

ADVANTAGES

Because the electromechanical system lacks the glamor of some of the more sophisticated systems to be described later, its advantages are often overlooked. The principle of operation is simple, and the circuits have few components. This leads to a highly reliable system. If properly installed and maintained with redundant hidden switches, the system can provide good protection. The electromechanical system is an excellent backup system for a more sophisticated system. The fact that it can be seen and easily identified will tend to discourage amateur burglars and vandals. The skilled burglar, on the other hand, may manage to thwart the system and, feeling that he is safe, walk right into a more advanced system that will alert the police.

Another advantage of this system is that it makes an excellent holdup alarm. In this use, during normal business hours, the regular protective circuit is disconnected. In its place is connected a loop containing switches that are hidden at various locations. If a holdup man pulls a gun in the area, one of the employees may manage to trip one of the hidden switches.

An ingenious holdup switch that may be thrown by the holdup man himself is sometimes used. This switch is part of the weight that normally rests on the money in a cash drawer. Whenever the weight is lifted more than a normal amount, the switch will be thrown. Usually a holdup man is in a hurry and will throw the switch when taking money out of the drawer.

CHAPTER 3

Photoelectric Detectors

The photoelectric detection system uses a beam of light to detect the presence of intruders. Fig. 3-1 shows a very simple system of this type. Normally, a beam of light from the light source is focused on the photocell. This causes the photocell to develop a small voltage that is amplified and used to energize a relay that carries alarm contacts. When light is shining on the photocell, the alarm contacts are held open. When the light beam is intercepted by an intruder, the photocell is darkened and its output voltage drops. As a result, the relay is de-energized and the alarm contacts close. If a separate battery is used for the alarm circuit, this system will fail safe; that is, the alarm will be set off if the light source fails or if the power lines to the system are cut.

This simple system has many limitations and is rarely used. The fact that the light beam is visible is an open invitation to the intruder to crawl under or leap over it. Even this is not necessary

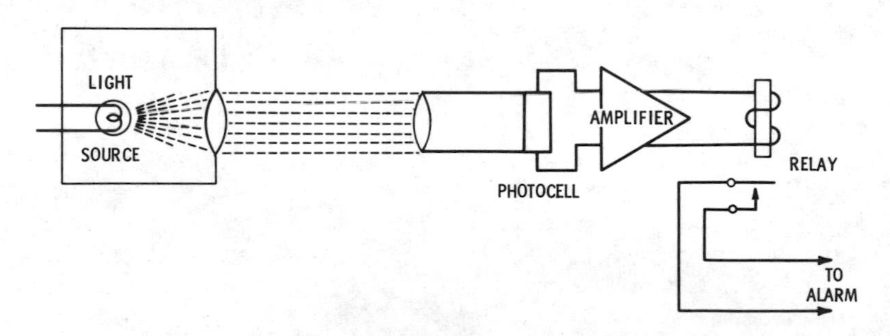

Fig. 3-1. Basic photoelectric system.

with the simple system shown, because it can be foiled by merely shining an ordinary flashlight into the photocell.

Many of the limitations of the simple photoelectric intrusion detector are overcome in more advanced systems. Most systems in use today use either infrared or ultraviolet light sources with filters to remove as much visible light as possible. In some of these systems there is still a faint red glow that can be spotted by a burglar. Well-designed systems, however, have almost invisible beams.

MODULATED LIGHT BEAMS

It was pointed out previously that the simple photoelectric system could be foiled by merely shining a light into the photocell. This serious limitation is overcome by using a modulated light beam. The system is arranged so that it will respond only to light with the proper modulation and thus will not be fooled by another light source.

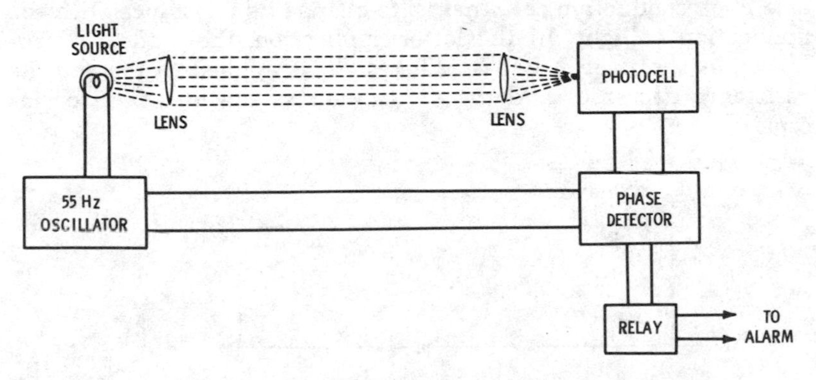

Fig. 3-2. Modulated light-beam intrusion alarm.

A block diagram of a modulated light system is shown in Fig. 3-2. Here, the light source is driven by a low-frequency oscillator so that the beam that it emits will be modulated at the frequency of the oscillator. The oscillator also provides a reference signal to one side of a phase-detector circuit. The other side of the phase detector is connected to the photocell. When the signal from the photocell is in phase with the reference signal from the oscillator, the dc output of the phase detector is maximum. An RC circuit connected to the output of the photocell compensates for inherent phase shifts in the system. If the light beam is interrupted, only one side of the phase detector is excited and its average dc output will be zero, de-energizing the relay and setting off the alarm. In a similar manner, the alarm will be set off if a steady light, or even a modulated light that is not in phase, is applied to the photocell. The frequency of the oscillator is usually low—one manufacturer uses 55 Hz. Multiples or submultiples of the line frequency are not used to prevent foiling the alarm with a light source synchronized to the power line.

Systems that combine the modulated-beam principle with an invisible infrared source are quite difficult to foil.

PHOTOJUNCTION DIODES AND PHOTOTRANSISTORS

Although there are still many photoelectric intrusion detectors in use that utilize vacuum tubes and vacuum phototubes, modern systems are all solid-state. The simplest solid-state photoelectric device is the photovoltaic cell, such as the familiar solar cell. This device generates an output voltage whenever light shines on it.

The photojunction diode may be used either as a photovoltaic cell, that is, to generate a voltage with the application of light; or as a photoconductive cell, that is, to change its resistance with the application of light. In the latter application, the diode is back-biased as shown in Fig. 3-3. When light is applied to the cell, its resistance decreases and the voltage across the load resistor increases.

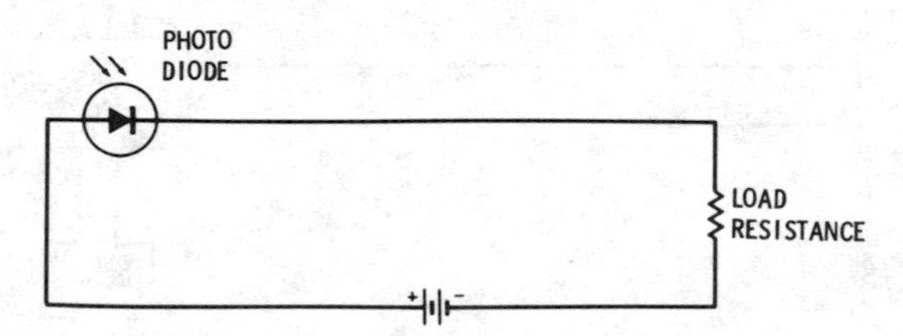

Fig. 3-3. Photodiode used as a photoconductive cell.

A comparatively new device that is finding use in photoelectric alarm systems is the phototransistor. This device is similar in operation to an ordinary transistor in some respects. Fig. 3-4 shows a circuit diagram of a phototransistor connected to a load. Note that there is no physical connection to the base, which is photosensitive. Normally, with the bias voltages shown, the collector current is very small. However, when light falls on the base p-region, holes are generated in sufficient quantity to allow somewhat less than 10 mA of current. This is enough to operate a relay directly. The phototransistor provides a lot of current amplification. That is, the small potential developed in the base region by the light controls a comparatively large collector current.

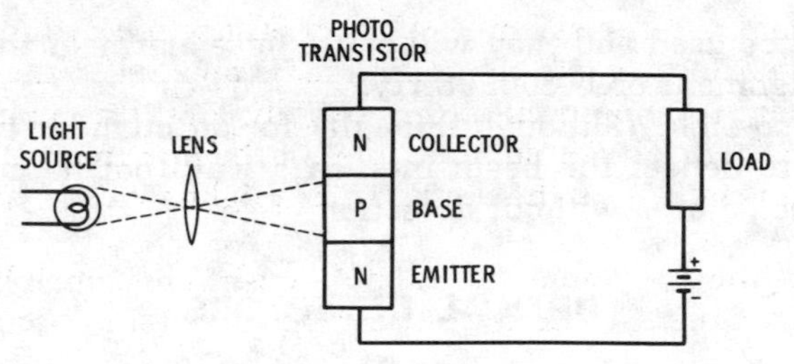

Fig. 3-4. Phototransistor detection circuit.

Applications

By its very nature, the light beam will travel only in a straight line. This makes it well suited for detecting entrance to an area where there is open space and nothing will normally interfere with the beam. Coverage of larger areas can be provided by an arrangement of mirrors to deflect the light beam as shown in Fig. 3-5. With this arrangement, an intruder will be detected either when he enters an area or when he starts to move around.

Limitations

The principal limitations of the photoelectric system are:

1. It is difficult to apply to areas where there are no long, straight paths for the light beam. In such cases, many mirrors

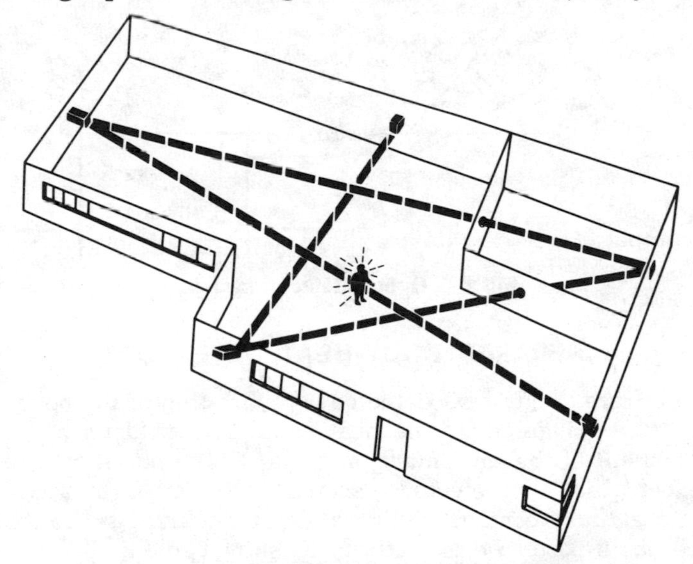
Fig. 3-5. Light beam deflected throughout area by mirrors.

must be used and they will cause false alarms if they get out of alignment or become dirty.

2. It is possible (although difficult) for an intruder to use mirrors to deflect the beam in such a way that he can enter a protected area without detection.

OPTICAL DETECTORS

Fig. 3-6 shows a photoelectric detector that does not depend on a light source for its operation. This system actually measures the ambient light in an area and reacts to any sudden changes. In operation, the photocell is focused directly on the area to be protected. Very slow changes in ambient light have little effect because they are averaged out in the RC coupling circuit. However, any change in level that occurs more rapidly—either from the presence of additional light or from the decrease in light caused by an intruder standing in front of the protected object—will trip the alarm. The sensitivity of a system of this sort must be adjusted for the particular application. It can be adjusted so that lighting a single match in a dark room will set off the alarm.

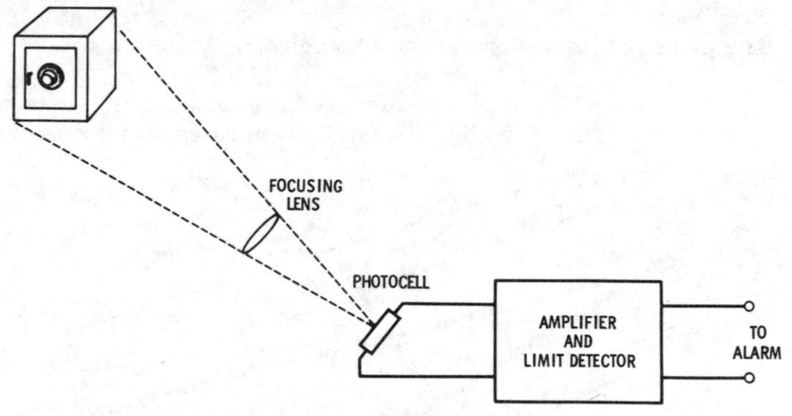

Fig. 3-6. Optical detector system.

INFRARED BODY-HEAT DETECTORS

The infrared body-heat detector is a variation of the optical detector. It is triggered by the heat from an intruder's body. The arrangement of the equipment is similar to the one shown in Fig. 3-6 except that the detector responds to infrared radiation. The detector actually consists of several detectors arranged so that the device has a pattern of sensitivity as shown in Fig. 3-7. The detectors are adjusted so that they are most sensitive to radiation

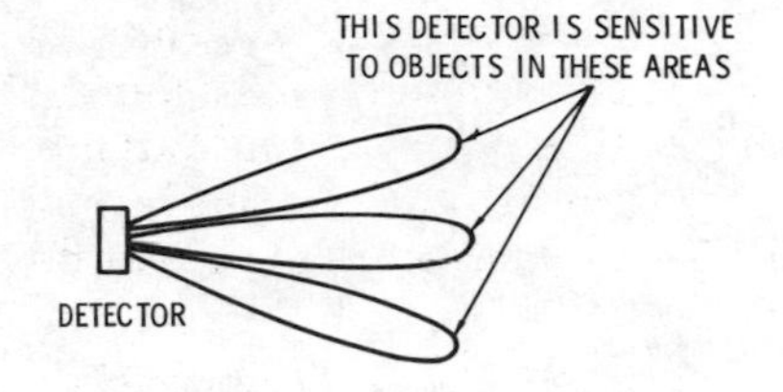

Fig. 3-7. Sensitivity pattern of an infrared body-heat detector.

from a source having a temperature of 98.6°F, the normal temperature of the human body.

If the temperature of the entire room should vary up or down, the detector would not respond to the change. But if an object (such as an intruder) having a temperature approximately equal to body temperature were to pass through the pattern from a sensitive area to a nonsensitive area, the device would detect the difference in radiation and initiate an alarm.

Infrared body-heat detectors are quite sensitive and not as easy to foil as they might seem. Superficially it would seem that all an intruder would have to do to frustrate the alarm would be to completely cover himself with something like a sheet. Theoretically this would work, but the sheet would have to be at the same temperature as the background temperature in the room. This would be hard to accomplish. The most effective way to frustrate an alarm of this type is for the intruder to move very slowly through the area.

CHAPTER 4

Ultrasonic Intrusion Detectors

The ultrasonic intrusion detector uses a beam of ultrasonic energy to detect the presence of an intruder. Ultrasonic energy is merely sound waves that have a frequency too high to be detected by the human ear—usually in the range of 20 to 50 kilohertz. Since ultrasonic energy cannot be heard, seen, or felt, the system has the obvious advantage of not being easily detected.

PRINCIPLE OF OPERATION

Fig. 4-1 shows an ultrasonic detector unit, and Fig. 4-2 illustrates a typical installation to protect a small room. Ultrasonic waves from the transmitter reach all parts of the room and are reflected many times before arriving at the receiver. Most of the energy reaching the receiver comes directly from the transmitter, but a small portion of the reflected energy also reaches the receiver. The direct and reflected waves combine in the receiver and tend to either reinforce or cancel each other, depending on their phase relationship.

As long as nothing moves in the protected area, the signal at the receiver will be constant. If an intruder moves about in the protected room, the reflected signal will change in both amplitude and phase. This change in the reflected signal results in amplitude modulation of the signal at the receiver. This modulation is detected, and the alarm is initiated.

Fig. 4-1. Typical ultrasonic intrusion detector.

ULTRASONIC WAVES

The direct and reflected waves combine in the room to set up a "standing wave" of ultrasonic energy. The generation of the standing wave can easily be understood by comparing it to the generation of waves of motion along a rope. Suppose that the far end of a rope is not attached to anything and the rope is given a shake at the near end. A single wave of motion will travel along the rope as shown in Fig. 4-3A. If the shaking is continued rhythmically, a continuous traveling wave will be sent down the rope, as in Fig. 4-3B. Suppose that the far end of the rope is securely fastened, as in Fig. 4-3C, and it is given a single shake. As shown in Fig. 4-3C,

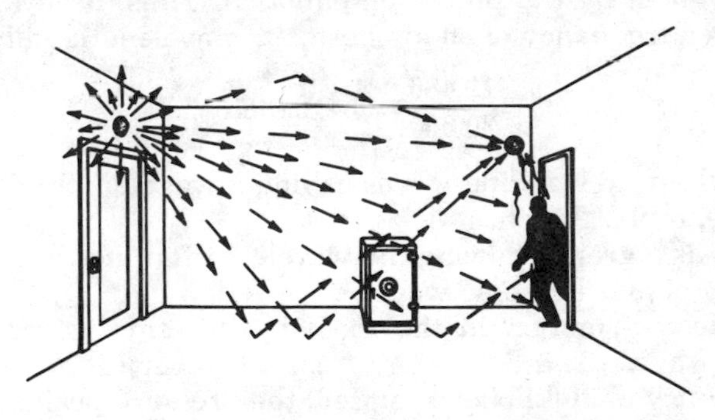

Fig. 4-2. Ultrasonic intrusion alarm installed in office.

the single wave will now travel down the rope to the end, be reflected, and travel back to the starting end.

If the rope is shaken rhythmically, its motion will be the sum of the motion due to the direct wave and the motion due to the reflected wave. Since the two waves travel at the same speed, one going toward the far end and one coming back, the net result is that the waves do not move at all. There is a standing wave on the rope. The rope vibrates between nodes which do not move at all.

At every point along the rope, the actual displacement is due to the combined effects of the direct and the reflected waves. The actual displacement at any point can be found by algebraically adding the displacement due to the direct wave alone and the displacement due to the reflected wave alone. The nodes occur when the algebraic sum of the two waves is zero. Ultrasonic direct and reflected waves combine in the same way to provide standing waves in a room.

The length of the standing waves depends on the wavelength, and hence on the frequency of the ultrasonic energy used in a particular installation. Since the velocity of sound in air is approximately 1040 feet per second, or 12,480 inches per second, we can compute the wavelength from the frequency, using the formula:

$$\lambda = \frac{V_s}{f_s}$$

where,

λ is the wavelength of sound in inches,
V_s is the velocity of sound in inches per second,
f_s is the frequency of sound in hertz.

Since the standing waves are produced by the interaction of the direct and the reflected waves, they will have a wavelength one-half that given by the preceding equation. Thus, if an ultrasonic signal has a frequency of 30 kilohertz, its wavelength will be

$$\frac{12{,}480}{30{,}000} = 0.416 \text{ inch}$$

and it will set up standing waves having a wavelength of one-half this value, or 0.208 inch.

Fig. 4-4 is a graph showing the wavelength in inches versus the frequency of an ultrasonic wave.

The effect of motion in the room is the same as moving the standing waves toward or away from the receiver. For example, if a frequency of 30 kilohertz is used, the pressure peaks are separated by 0.2 inch. Thus, if a reflecting surface moves toward the

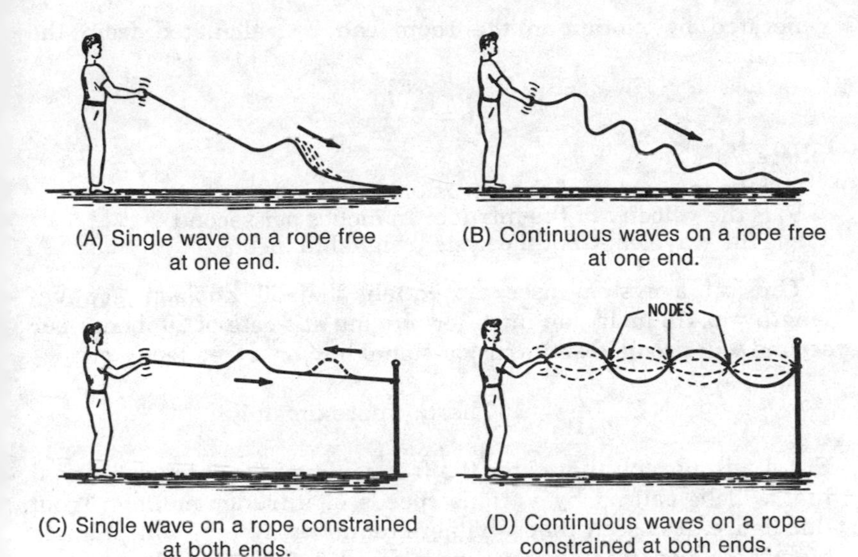

(A) Single wave on a rope free at one end.

(B) Continuous waves on a rope free at one end.

(C) Single wave on a rope constrained at both ends.

(D) Continuous waves on a rope constrained at both ends.

Fig. 4-3. Wave motions of a rope.

receiver at a rate of 0.2 inch per second, the signal at the receiver will appear to be modulated at a frequency of 1 hertz.

The ultrasonic alarm must be sensitive enough to detect small motions in the protected area but not be affected by changes caused by air current. The frequency of the "intrusion signal"

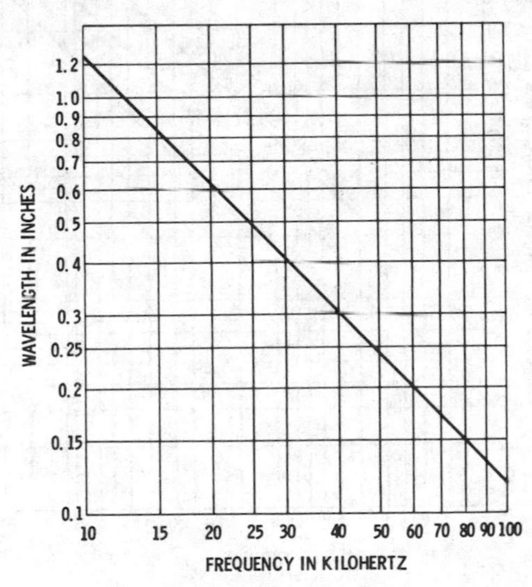

Fig. 4-4. Wavelength versus frequency graph of an ultrasonic signal.

generated by motion in the room can be calculated from the formula:

$$f_i = 2\frac{V_i}{\lambda}$$

where,

f_i is the frequency of the intrusion signal in hertz,
V_i is the velocity of the intruder in inches per second,
λ is the wavelength of the system signal in inches.

Thus, if a system uses a frequency of 30 kilohertz (wavelength = 0.416 inch), an intruder moving at a rate of 20 inches per second will produce an intrusion signal having a frequency of

$$2\frac{20}{0.416} = 100 \text{ hertz, approximately.}$$

Fig. 4-5 is a graph that gives the frequency of the intrusion signal that will be caused by various speeds of intruder motion. From this figure, it is seen that a higher-frequency system will produce higher-frequency intrusion signals for the same rate of motion of the intruder. Thus, higher-frequency systems are more sensitive, but on the other hand, higher frequency transducers are usually

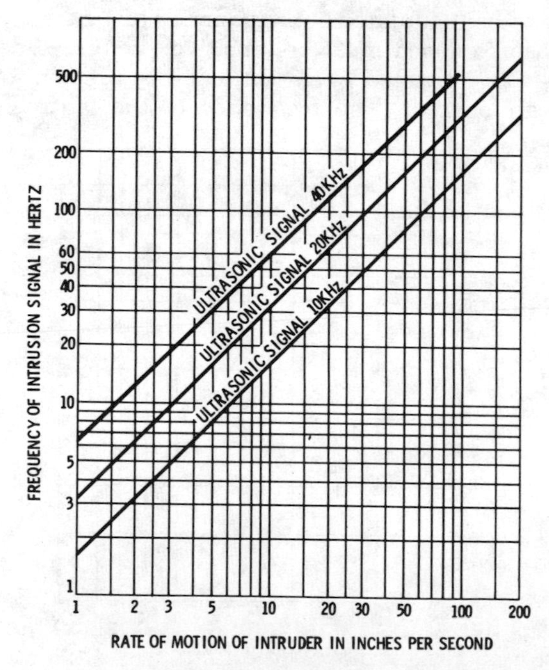

Fig. 4-5. Relationship of motion of intruder to frequency of an intrusion signal.

more expensive, and less sensitive. In practice, frequencies from about 20 to 50 kilohertz are commonly used.

The preceding analysis of the intrusion signal is based on the intruder moving either directly toward or directly away from the transducer. In actual practice, this analysis is very close to what is actually experienced. The ultrasonic wave is reflected so many times from one surface to another in the protected area that motion in any direction will cause an intrusion signal. Therefore, the curves of Figs. 4-4 and 4-5 are adequate for planning an actual system.

The effectiveness of an ultrasonic system depends on the energy being reflected many times in the protected area so that standing waves will be set up everywhere in the room. Hard surfaces, such as walls, desks, and file cabinets, are good reflectors of sound. Soft materials, such as carpets, draperies, and clothing, are not good reflectors of sound. Therefore, a small area with hard walls and reflecting surfaces will require fewer transducers than an office with wall-to-wall carpeting and many draperies. An area that is completely filled with soft material would probably be better protected by some other type of system.

TRANSDUCERS

In order to get ultrasonic energy into a room, a transducer that converts electrical signals into sound pressure waves is used. Since the function of the transmitting transducer is exactly the same as that of a speaker in a radio, it might appear that an ordinary speaker might be used as a transducer. Unfortunately, most speakers do not perform well at ultrasonic frequencies, so special transducers made for the purpose are used. These transducers work on the same principles as radio speakers, but they have small, stiff diaphragms so that they will operate well at ultrasonic frequencies. They more closely resemble a microphone in construction than a speaker. In fact, an ordinary crystal microphone cartridge may be used for a transducer in a home-constructed system.

The receiving transducer has the opposite function: it must convert sound pressure waves into electrical signals. Its function is the same as that of a microphone in a sound system. In actual practice, identical transducers are often used for both transmitting and receiving.

Just as regular sound microphones are made to have different directional patterns, ultrasonic transducers are available with different patterns. Highly directional units are available for use in long, narrow areas such as corridors. Omnidirectional units are

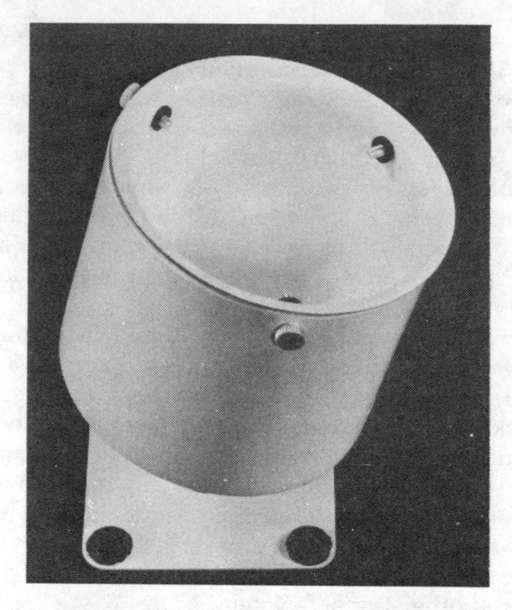

Fig. 4-6. Typical ultrasonic transducer.

used to protect rectangular areas. A typical ultrasonic transducer with cover is shown in Fig. 4-6.

GENERATING ULTRASONIC ENERGY

The transmitter of an ultrasonic intrusion alarm is very simple, consisting merely of an oscillator operating at the desired frequency and a transducer. In modern systems, the oscillator is usually a small solid-state unit that can be mounted inside the

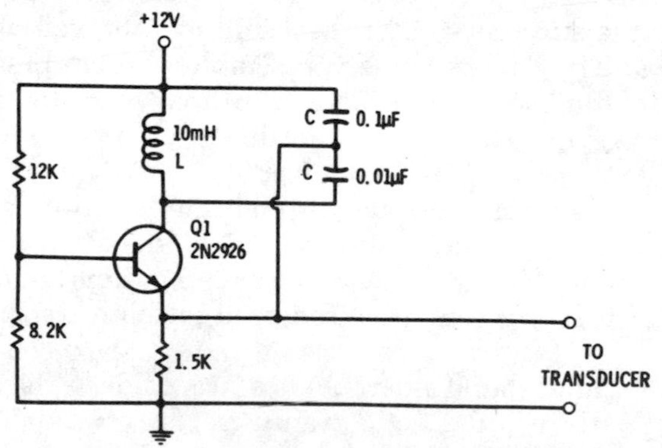

Fig. 4-7. Ultrasonic oscillator circuit.

transducer case. A typical oscillator circuit is shown in Fig. 4-7. This is a simple emitter-coupled oscillator. The frequency is determined by the values of L and C in the circuit. Usually, the inductance is variable so that the frequency can be set at the desired value.

Since the transmitter portion of the system is so simple, it is often mounted together with its transducer in the same case as the receiver. The unit in Fig. 4-1 uses this type of construction.

RECEIVER

Fig. 4-8 shows a block diagram of a typical receiver. The first element is the transducer, which converts the ultrasonic energy to an electrical signal at the ultrasonic frequency. The input circuit is usually tuned to the frequency of the transmitter so that it will pass only frequencies close to the frequency of operation. This limiting of the bandwidth of the system is very effective in minimizing false alarms which might otherwise be caused by extraneous sounds.

The signal is then amplified and applied to a detector. A gain control on the amplifier sets the sensitivity of the system. It should be remembered that at this point in the circuit the signal is a high-frequency audio voltage. If nothing is moving in the protected area, the amplitude of this signal will be constant, so there will be no output from the detector. When an intruder moves in the protected area, the ultrasonic signal is modulated by the low-frequency intrusion signal. Since this intrusion signal modulates the ultrasonic signal, the amplifier does not have to pass the low-frequency signal any more than the rf stages of a radio have to pass the audio signals.

When the ultrasonic signal is modulated, the intrusion signal will appear at the output of the detector. The frequency of this signal depends on the rate of motion of the intruder, as explained before. The amplitude of the signal depends on the obstructing area of the intruder.

The detector is usually followed by a high-pass filter that will eliminate very-low-frequency intrusion signals. This prevents the

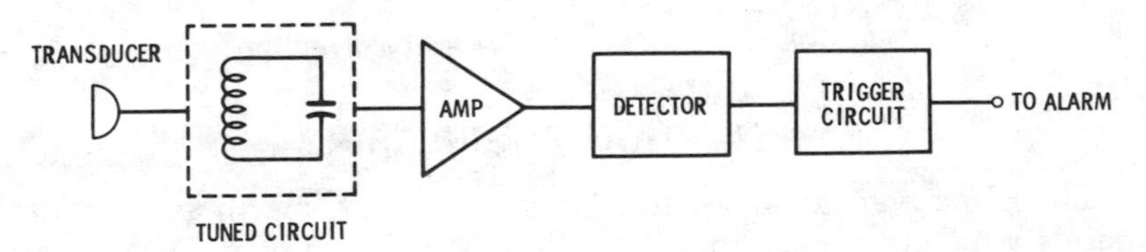

Fig. 4-8. Block diagram of an ultrasonic receiver.

system from being triggered by normal air currents in the protected area. The adjustment of this filter is a compromise between a condition that will detect the slowest moving intruder and one that will be triggered by air currents.

The filtered intrusion signal is applied to a trigger circuit like that shown in Fig. 4-9. The first two stages form a Schmitt trigger circuit that is really a sort of regenerative switch. Initially, transistor Q1 is shut off and Q2 is conducting. When the input signal exceeds a certain level, determined by the setting of R1, Q1 will start to conduct and Q2 will be shut off. Thus, this circuit will convert the intrusion signal into a square wave signal whenever it exceeds the threshold level set by R1. Below this signal level, there is no output. The output of the Schmitt trigger drives Q3, which drives the relay, which in turn actuates the alarm.

It is quite common to include a time delay in the circuit so that the alarm will not go off until the intrusion signal has been present for 10 seconds, or even longer. This arrangement eliminates false alarms that might be caused by radio-frequency interference, curtains occasionally moving in the breeze, and transients on the power line.

Since the magnitude of the intrusion signal depends on the area of the intruder, the system can be adjusted so that it will set off the alarm if even a small man moves very slowly in the area, but will not be affected by the motion of insects, mice, etc.

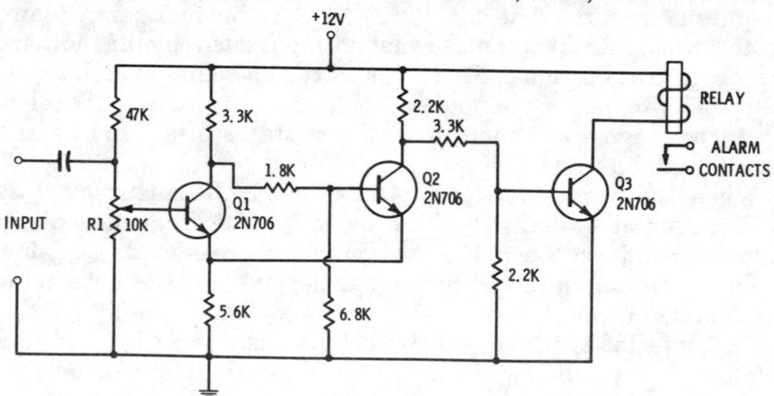

Fig. 4-9. Trigger circuit for an intrusion alarm.

APPLICATION ENGINEERING

The effectiveness of an ultrasonic intrusion alarm system depends very greatly on how well the particular application is engineered. The system must be made as foolproof as possible and yet produce a minimum number of false alarms. This is not an easy task. In general, the more sensitive a system is, the more likely it is to be triggered by some extraneous condition.

The first consideration in an installation is selection of an optimum location for the transducers. When the area to be protected is small, fairly open, and rectangular in shape, the problem is relatively easy. The transducers should not be pointed directly at anything that might move or that has moving parts, such as electric fans. They should not be pointed directly through areas where air convection currents are apt to be strong, such as in front of air conditioners or over radiators. Otherwise, such an installation is straightforward.

Where the area to be protected is broken up by desks, cabinets, and other furniture, selecting a transducer location is often difficult. Even locations that appear to be adequate for complete coverage may leave large "blind" areas that are not sufficiently covered by the ultrasonic energy. The usual procedure in such a case is to select what appears to be the best location for the transducers and then test the system by walking through the protected area slowly. Another important part of the test is to determine if anything that normally moves or produces sound will trigger the circuit and cause a false alarm. Of course, it does not matter if something that should not be operated after the place is closed causes an alarm. In fact, this provides a measure of additional protection. For example, in an installation in a cocktail lounge, the cash register apparently produced some ultrasonic energy when it was operated. As a result, the alarm was triggered whenever the cash register was operated. This was no problem because the cash register should not have been opened after the place was closed. On the other hand, in another installation, a valve on a steam radiator generated ultrasonic energy at all times. This was the source of many annoying false alarms until the transducers were relocated.

There are a few general considerations that will help in finding the best location for transducers.

1. Use a transducer with the proper pattern. If there are many normally moving objects in the area, it is best to use a directional transducer.
2. In small areas, mount the transducers near the corners of the area, preferably on the ceiling.
3. Mount the transducers on a surface that is free from vibration.
4. If the ceiling are more than 12 feet high, mount the transducers on the walls. Avoid walls that can be drilled through from the outside.

5. Keep the transducers at least 10 feet from objects that can emit high pitched sounds, such as telephone bells, radiator valves, and steam pipes.
6. Locate transducers as far as possible from moving objects such as drapes, curtains, fans, and machinery.

After the transducers have been placed in the optimum location, the system must be adjusted or "tuned-up." This consists of adjusting the sensitivity controls to get optimum performance. Many systems also have time-delay controls and low-frequency response adjustments. These must be carefully adjusted in accordance with the manufacturer's instructions.

The final evaluation of the system is the "walk-through" test. After the system is installed and adjusted, the technician should try to fool it in any way possible.

ADVANTAGES AND LIMITATIONS

The ultrasonic system has many advantages over the systems described in previous chapters. One of its big advantages is that it is not easy to identify. This makes it difficult for a burglar to attempt to foil it. Unlike the perimeter protection systems, it will spot the "stay-behind" the minute he starts to move about in the area.

There are some limitations to the ultrasonic system. Is it not suitable for use in areas that contain equipment that inherently makes high-pitched noises. Since it is operated by ultrasonic energy, it can be jammed or triggered by ultrasonic energy. Because the system depends on reflection for its operation, it is hard to adapt to areas that contain large amounts of sound-absorbing material. For this same reason, it is not suitable for protection of completely open areas such as storage yards.

Many of the limitations and disadvantages that have been experienced in actual ultrasonic system installations arise because the application has not been properly engineered. In all but the simplest applications, it is not advisable simply to buy a system and put it in. Careful planning and engineering are needed to ensure an effective system.

CHAPTER 5

Microwave Intrusion Detectors

The microwave intrusion detector shown in Fig. 5-1 operates similarly to the ultrasonic intrusion detector described in the previous chapter. One principal difference is that the ultrasonic system utilizes sound pressure waves in air, whereas the microwave system uses very short radio waves. The microwave system is often called a radar alarm because it is actually a form of Doppler radar.

PRINCIPLE OF OPERATION

The principle of operation of the microwave intrusion detector is exactly the same principle by which a moving airplane causes the picture on a television receiver to flutter on the screen. As shown in Fig. 5-2, the signal from a television broadcasting station reaches a particular receiver through different paths. One is a direct path from the transmitter. In the other path, the signal is reflected from the plane before reaching the receiver. The received signal is the algebraic sum of the signals from the two paths. If the plane were not moving, the signal strength at the receiver would be constant. When the plane moves, however, the phase of the reflected signal changes with respect to that of the direct signal. It alternately reinforces and cancels the direct signal. The result is that the received signal is amplitude-modulated at a frequency that depends on the speed and direction of the plane's motion, causing the flutter of the picture on the receiver.

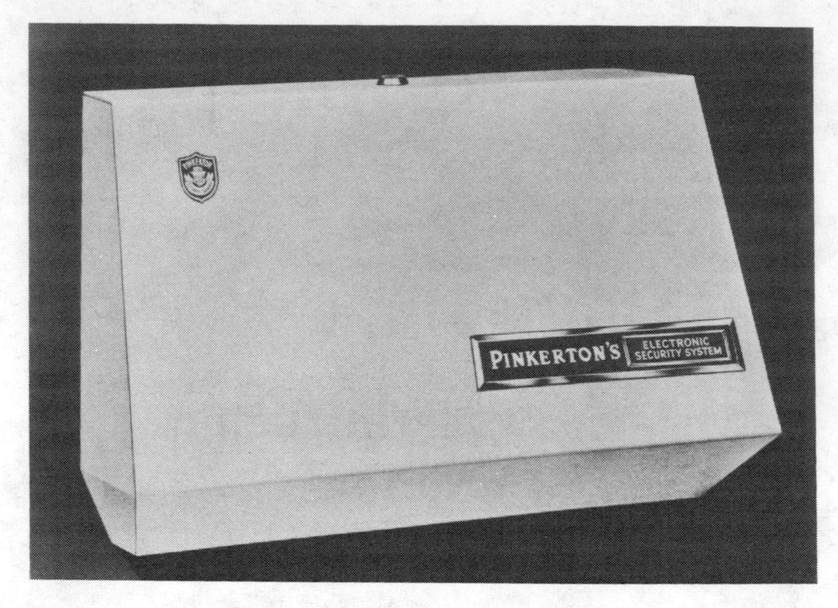

Fig. 5-1. Microwave intrusion detector.

In a typical microwave intrusion detector, microwave energy from the transmitter fills the protected area and sets up standing waves in much the same way as in the ultrasonic system. Two

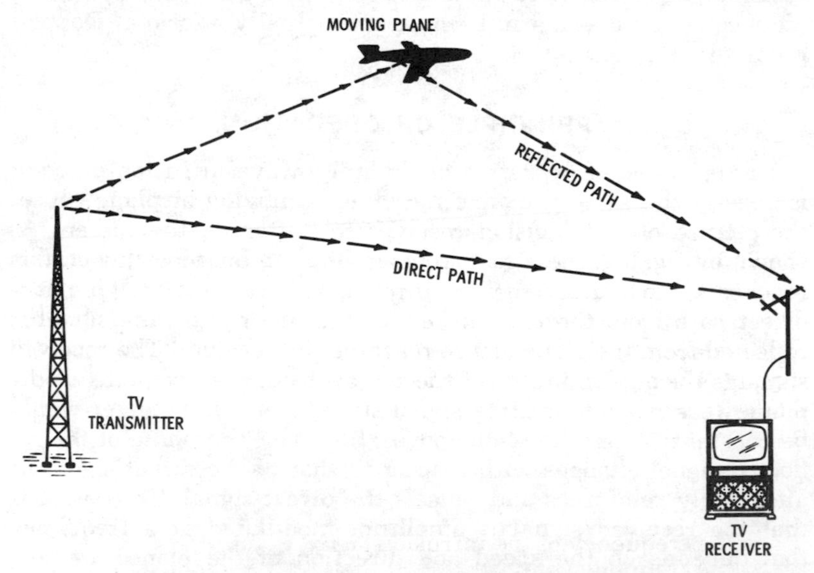

Fig. 5-2. Principle of microwave intrusion detector.

separate signals arrive at the receiver. One is a direct signal from the transmitter. This is received either by radiation from the transmitter, or often by direct connection. The other signal is the complex signal reflected from many different surfaces in the protected area. As long as nothing moves that will reflect a signal, the signal strength at the receiver is constant and is the algebraic sum of the direct and reflected signals. When an intruder enters the area, he upsets the standing wave pattern and causes the amplitude and phase of the received signal to vary at a rate that depends on how fast he is moving. This has the effect of amplitude-modulating the received signal by a low-frequency intrusion signal.

FREQUENCIES USED

The wavelength of a microwave signal depends on the speed of propagation of radio waves and the frequency of the signal. The wavelength is given by the formula:

$$\lambda = \frac{300}{f}$$

where,

λ is the wavelength in meters,
f is the frequency in megahertz.

To find wavelength in inches, use the formula:

$$\lambda = \frac{11{,}811}{f}$$

where,

λ is the wavelength in inches,
f is the frequency in megahertz.

Various commercial microwave intrusion detectors operate anywhere from 400 to 25,000 megahertz. Fig. 5-3 gives the wavelength of microwave signals in inches for various frequencies. From this figure, we see that a signal of 10,000 megahertz will have a wavelength of 1.2 inches.

The frequency of the intrusion signal depends on the rate of motion of the intruder and the frequency of operation of the system. The frequency is given by:

$$f_i = \frac{2f_s \times S_i}{11{,}811}$$

where,

f_i is the frequency of the intrusion signal in hertz,
f_s is the frequency of operation of the system in megahertz,
S_i is the rate of motion of the intruder in inches per second.

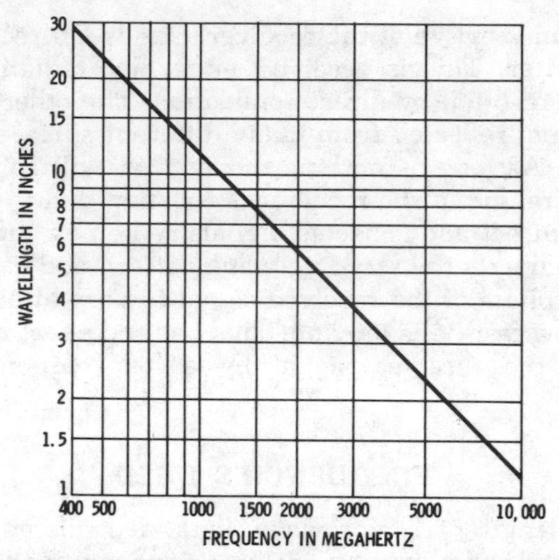

Fig. 5-3. Frequency versus wavelength of microwave signals.

The factor of 2 is needed in the foregoing equation because the wavelength of the standing wave is one-half that of the system signal.

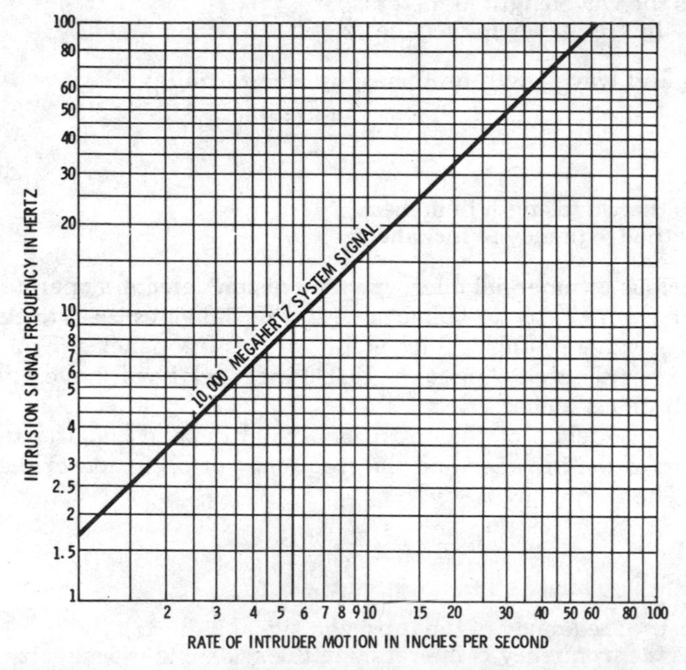

Fig. 5-4. Intrusion signal frequency versus speed of intruder motion.

Thus, in a 10,000-megahertz system, an intruder moving at a rate of 5 inches per second will generate an intrusion signal of 8.46 hertz. Fig. 5-4 gives the frequency of the intrusion signal for various rates of intruder motion in a 10,000-megahertz system. Since the frequency of the intrusion signal is directly proportional to the system frequency, if the system frequency were cut in half, the frequency of the intrusion signal would also be cut in half. Thus, higher-frequency systems tend to be more sensitive to slower rates of intruder motion.

PROPERTIES OF MICROWAVES

Although the principle of operation of the microwave system is somewhat similar to that of the ultrasonic system, the two systems differ in many important respects. The ultrasonic system uses sound waves in air. These waves are actually air-pressure waves. In general, they are reflected by hard surfaces and are absorbed by soft surfaces. Microwaves are actually high-frequency radio waves. They are reflected well by metal surfaces, but most interior building materials such as wood, glass, and wallboard are almost transparent to microwaves. The signal will go right through them. This is both an advantage and a limitation. As an advantage, it permits protecting several rooms or offices with one system—something that could not be done with an ultrasonic system. As a limitation, it is sometimes difficult to contain the microwave signal in the protected area. Since the signals will pass through glass easily, the signals may be reflected by a moving automobile outside, causing a false alarm. More will be said about this later under application engineering.

Since microwaves are radio waves, they are usually not affected by air currents such as would be produced by heaters or air conditioners. Of course, the microwave system is not affected at all by extraneous noises which could easily occur in the protected area.

TYPICAL SYSTEMS

The state of the art in solid-state devices and circuits is changing rapidly. A few years ago, all microwave intrusion alarms used vacuum tubes because transistors were not available to operate at these frequencies. Even now, vacuum-tube units using klystron tubes are available, but the trend is more and more toward solid-state units. In the following paragraphs a klystron unit is described, followed by descriptions of a few typical solid-state units.

Antennas

In general, for an antenna to have a directional pattern, its longest dimension must be greater than one wavelength. For this reason, higher-frequency systems usually employ directional antennas to concentrate the microwave energy in the desired area. At a frequency of 400 megahertz, one wavelength is almost 30 inches, so directional antennas are not practical. These systems usually employ a vertical antenna one-quarter wavelength long. Naturally, this type of antenna tends to radiate equally well in all directions, except from the end.

A Typical 10,000-Megahertz System

Fig. 5-5 shows a typical microwave system of the type that operates at frequencies between 1000 and 10,000 megahertz. The signal is generated by a vacuum-tube klystron oscillator.

The klystron oscillator circuit is very simple. The frequency of operation is determined by the klystron itself, so the only electronic circuitry required is a regulated power supply. The rf connection from the resonant cavity of the tube couples the microwave energy directly to the antenna. Systems of this type usually have an rf power output well under 1 watt and can provide a signal strong enough to protect an area up to about 150 feet long.

At the frequency of 10,000 megahertz, one wavelength of the signal is only about 1.2 inches. Therefore, it is possible to build an antenna having dimensions of several wavelengths that still is not unduly large physically. The horn antenna shown in Fig. 5-6 is the most commonly used type. Horn antennas are available with a wide variety of different patterns as shown in Fig. 5-7.

Referring to Fig. 5-5, the signal generated by the klystron oscillator is fed to the antenna where it is radiated into the area to be

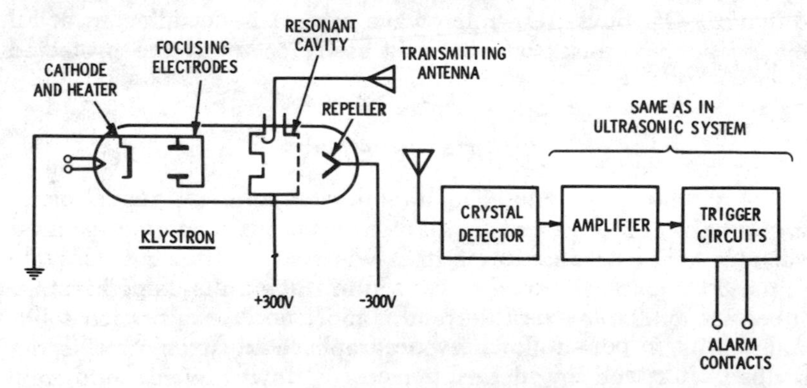

Fig. 5-5. Typical 10,000-megahertz system.

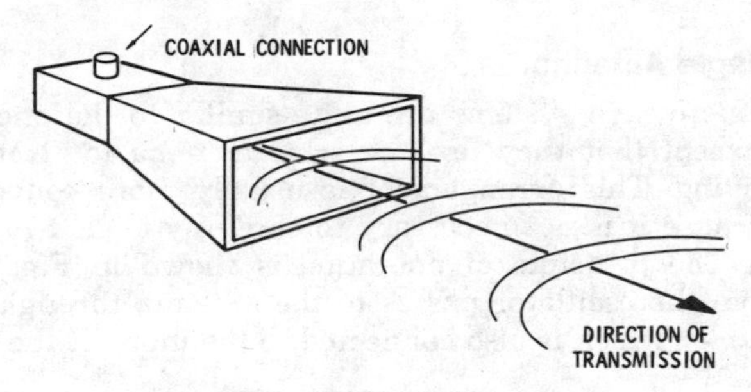

Fig. 5-6. Rectangular horn antenna.

protected. The receiver has an identical antenna pointed in the same direction, so it picks up any microwave reflections from objects in the area. A small portion of the signal from the transmitter reaches the receiver antenna directly in most installations. In some cases, the signal is fed directly to the receiver through a small coaxial cable.

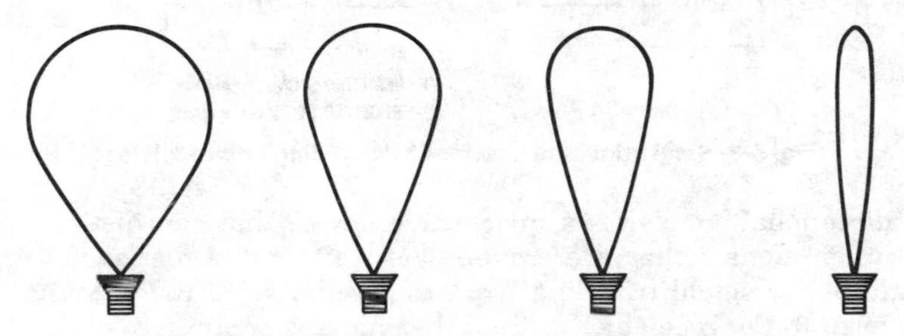

Fig. 5-7. Directional patterns of some horn antennas.

As long as nothing is moving in the protected area, the rf signal at the receiver antenna is constant. When an intruder moves through the protected area, he will change both the amplitude and phase of the reflected signal. This causes the signal at the receiver to be amplitude-modulated by the intrusion signal. In most microwave systems, the first stage of the receiver is a crystal detector that demodulates the signal so that its output is the low-frequency intrusion signal. The intrusion signal is amplified and applied to a trigger circuit in much the same way as it is in the ultrasonic system. Filters are sometimes used to filter out all signals except those that would be caused by human-like movements. Time-delay relays are usually provided so that the intrusion signal will have to persist for a few seconds before the alarm is set off. A gain control is provided in one of the amplifiers of some systems to permit setting the sensitivity of the complete system.

Use of Single Antenna

Some microwave systems are quite similar to the one just described except that they use the same antenna for transmitting and receiving. This arrangement is usually more convenient to install because it is not necessary to carefully align two separate antennas. The principle of operation is shown in Fig. 5-8. The signal from the oscillator passes to the antenna through a *directional coupler* which is also connected to the input of the receiver.

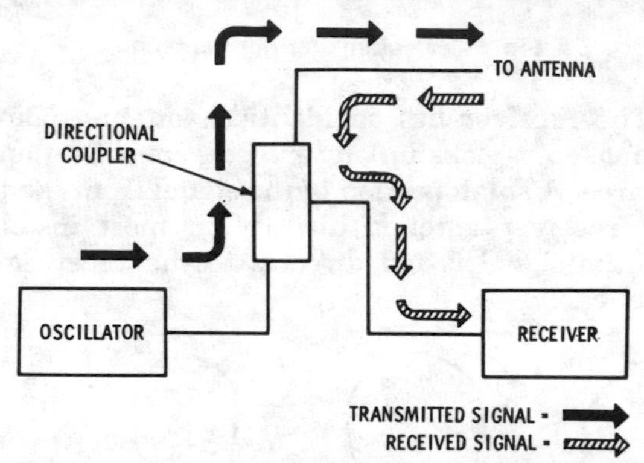

Fig. 5-8. Single antenna used for transmitting and receiving.

A directional coupler is a microwave device that can distinguish between signals that are traveling in different directions. Very little of the signal traveling from the oscillator to the antenna is coupled to the receiver. On the other hand, a comparatively large amount of the reflected signal that is traveling in the opposite direction is coupled to the receiver. This system is otherwise the same as that shown in Fig. 5-5.

TYPICAL SOLID-STATE SYSTEM

Fig. 5-9 shows the schematic diagram of a typical microwave oscillator used in intrusion alarms. This is a conventional emitter-coupled oscillator that can be used to at least 500 megahertz. At this frequency, the tank inductance, L1, does not have any turns. It is simply a rod about 1.5 inches long, connected directly from the collector of the transistor to ground. The output is tapped to the rod about 0.25 inch from the ground end. This circuit will produce an rf power output of about 10 milliwatts with the components shown. This is adequate for most intrusion alarms. The receiver would be identical to the receiver used with the klystron system described in the previous section.

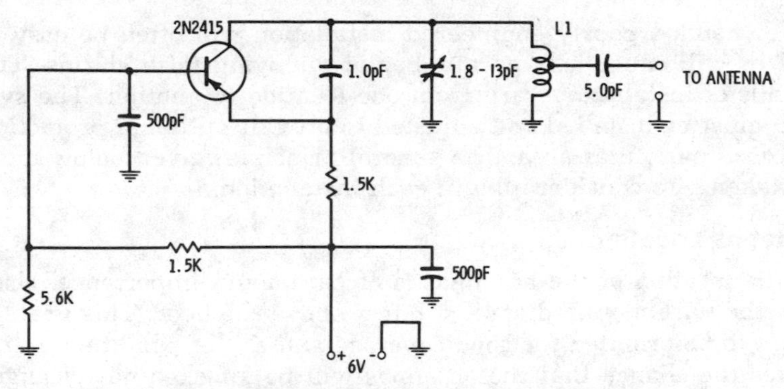

Fig. 5-9. Typical solid-state transmitter circuit.

Solid-state units that operate at higher frequencies are rapidly appearing on the market. In these units, the oscillator tank circuit is usually a cavity. Fig. 5-10 shows the diagram of a transistor cavity oscillator that will provide about 0.5 watt of power at frequencies over 1000 megahertz. The tank circuit is a rectangular cavity about 1 inch square by 1.75 inches long.

As more microwave transistors become available at lower prices, more solid-state systems can be expected.

APPLICATION ENGINEERING

The microwave intrusion detector will provide a very effective protection system if the application is properly engineered. On the

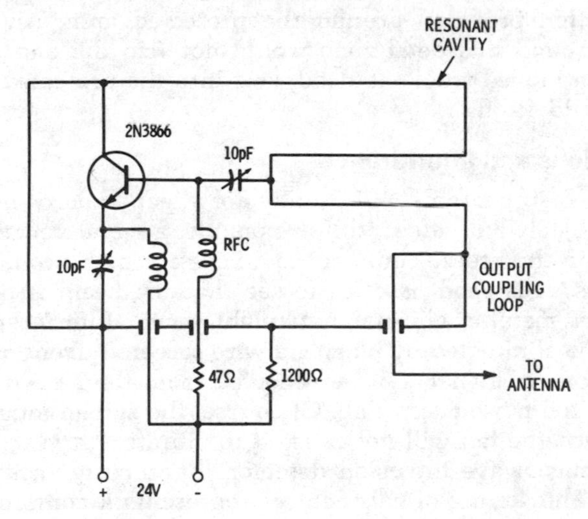

Fig. 5-10. Microwave oscillator using a transistor in a resonant cavity.

other hand, a poorly engineered installation will often be easy to foil or will cause a large number of annoying false alarms. The details of installation vary from one location to another. The system must be installed and adjusted to provide optimum protection for each particular area. The general principles given below must be taken into consideration in each installation.

Antenna Location

The location of the antennas is of paramount importance. Usually they are mounted at least 8 feet above the floor. This enables them to be oriented for good coverage and at the same time minimizes the chance that the antennas will be knocked out of alignment during normal business hours when the system is not operating. The antenna must be mounted on a surface that is free from vibration. As pointed out in an earlier section, the higher-frequency systems are well suited to directional antennas that permit concentrating the microwave energy in the desired area.

Since microwave energy will pass through most interior wall materials easily, it is possible to protect more than one room with a single system. The signal will pass through at least two plasterboard walls and even more plywood walls.

Most exterior wall materials such as brick, concrete block, and masonry are reasonably opaque to microwaves, but care must be taken to avoid wooden exterior walls and glass windows. The antennas must not be pointed at wooden doors, walls, or glass windows because reflections from objects passing by outside will cause false alarms. In installations where it is practical, a dead zone is often provided around the protected area. An intruder passing around this dead zone would not trip the alarm, but as soon as he moved from the dead zone into the protected area the alarm would go off.

Obstructions and Interference

Although air currents will usually not affect a microwave alarm, moving objects will affect it. For example, a metal venetian blind swinging in the breeze could cause a false alarm. So could moving fan blades. It is good practice to see that the beam is not aimed directly at metallic objects that might move. Interference from fans can be eliminated by placing a wire screen in front of the fan. The microwave signal will be reflected from the screen and will not "see" the fan blades at all. Of course, the screen must be firm enough that the fan will not cause it to vibrate.

Many microwave intrusion detectors have range or sensitivity controls. The setting of this control represents a compromise between providing adequate protection at the longest range from the system and avoiding false alarms because of too much sensitivity at close ranges. The range setting must be correlated with the location of the antennas. For example, it would not be wise to try to adjust the sensitivity of a system in such a way that it would be able to detect a man walking at one location and yet ignore a large moving object such as an elevator just behind a thin wall. This could be accomplished better by locating the antennas so that they would not "see" the elevator.

The fact that metal objects are opaque to microwaves means that objects such as desks, file cabinets, and storage cabinets can create "shadows" in the protected area. That is, there will be no microwave energy and hence no reflections from areas behind metal objects. Unless care is taken in properly positioning the antennas, there might be a shaded path along which a burglar could crawl without being detected.

Testing the Installation

Regardless of how much care has been exercised in planning the installation, the final proof of performance is an actual test. In testing a system, anything that might produce a false alarm must be tried. Fans must be turned on, venetian blinds waved back and forth as they would in a breeze, and loose doors vibrated. If the protected building is near a street or sidewalk, the system should be tested to make sure that people or vehicles passing by outside will not cause false alarms.

The test should also include all possible attempts to foil the system, such as walking into the protected area very slowly or trying to reach a critical area by crawling under the beam or finding a "shaded" path.

ADVANTAGES AND LIMITATIONS

The microwave intrusion detector ranks with the ultrasonic detector as one of the most effective means of providing specific area protection. A properly installed system is very difficult to foil. It has the advantage that two or more separate rooms can often be protected by a single system.

The microwave alarm is very effective in trapping the "stay-behind." The minute he starts to move in the protected area, he will trip the alarm.

The principal limitation of the microwave system is that it requires adequate installation engineering. The fact that many outside walls contain large windows which are easily penetrated by microwaves can lead to frequent false alarms in poorly engineered installations.

Another limitation that must be considered concerns the power from a microwave system. The power must be held to levels that are not dangerous to human beings; under no condition should the power density be allowed to exceed 10 milliwatts per square centimeter.

In some geographical locations, microwave systems are subject to interference from high-powered radars such as those used in air-traffic control and defense establishments. This possibility should be checked out when a microwave system is being installed. In most cases, the interference can be eliminated by proper orientation of the antennas. In installations on the same property as radar systems, much more care must be taken. The system chosen should not be susceptible to radiation at the frequencies of the radars in use.

FCC RULES AND REGULATIONS

Any device that radiates rf energy, such as a microwave intrusion detector, can both cause interference and be susceptible to it. For this reason, the Federal Communications Commission carefully regulates all such devices. In the FCC rules, a microwave intrusion alarm is called a *field disturbance sensor* because it operates on the basis of sensing a disturbance in an electromagnetic field. The rules covering field disturbance sensors will be found in Part 15 of the *FCC Rules and Regulations,* which is available from the U.S. Government Printing Office, Washington, D.C. 20402.

Before any field disturbance sensor can be operated, it must be "certificated" and labeled in accordance with the FCC rules.

In general, a field disturbance sensor may be operated on any frequency as long as it does not cause interference to any other device or equipment. Unless special frequencies are used, however, the radiation on any frequency, including both the fundamental frequency and the harmonics, must be less than 15 microvolts per meter at a distance in meters from the sensor given by:

$$\text{Distance} = \frac{\lambda}{2\pi}$$

where,

λ is the wavelength in meters.

In more familiar units, the distance in feet is given in terms of frequency by:

$$\text{Distance} = \frac{157{,}000}{\text{frequency in kHz}}$$

From an inspection of the foregoing equation, it is rather obvious that the radiation from a microwave alarm that meets these requirements will be rather seriously limited. This means that the receiving portion of the system must be very sensitive in order to detect an intruder at a distance. This high sensitivity in turn means that the receiver will be quite sensitive to interference from other radiating sources, and this interference may well cause false alarms.

Table 5-1. Special Frequencies for Microwave Alarms (Field Disturbance Sensors)

Operating Frequency (MHz)	Band Limits (MHz)	Allowable Field Strength at 30 Meters*
915	±13	50,000 μV/m
2450	±15	50,000 μV/m
5800	±15	50,000 μV/m
10,525	±25	250,000 μV/m
24,125	±50	250,000 μV/m

* 30 Meters = 98.4 Feet

To avoid this limitation, the FCC will allow much higher radiation if the alarm is operated on certain frequencies and the frequency is held constant. Table 5-1 lists these special frequencies and gives the frequency tolerances as well as the permissible radiation.

In order to obtain FCC certification, a complete series of tests and measurements must be performed by a competent engineer and submitted to the FCC. Once the device has been certificated, no changes can be made that will affect the amount of radiation from the device. When repairs are made, tests must be made to ensure that the device will still meet the prescribed standards.

CHAPTER 6

Proximity Detectors

The proximity detector, as its name implies, is a device that will trigger an alarm whenever an intruder comes close to it; the intruder does not have to actually touch any part of the system.

The proximity detector is unique among intrusion detectors because, in general, no two detectors are exactly alike. Other types of intrusion detectors are manufactured products, and the manufacturer has usually done a great deal of engineering to ensure that the detector will operate properly if it is properly installed. The proximity detector, on the other hand, is usually a wire of arbitrary length or, in some cases, the actual object that is to be protected, such as a safe or file cabinet.

Unfortunately, many proximity detectors are not properly engineered for a particular application, with the result that they operate poorly or cause a large number of annoying false alarms. This reputation is not justified because a properly engineered proximity system is very reliable.

To properly select and install a proximity detector, it is essential to understand the principle of operation of the device. The most critical part of the system is the sensing wire or the object selected to be protected. This choice is made at the time of installation rather than at the time of manufacture.

PRINCIPLE OF OPERATION

The sensing circuit of a proximity detector is actually a capacitor, but as we will see this capacitor differs considerably from the capacitors normally used in electronic circuits. A typical arrangement is shown in Fig. 6-1. When no intruder is in the vicinity of the system, the electric flux passes from the sensing wire to the ground. The air between the two points acts as the dielectric and has a dielectric constant of one (Fig. 6-1A).

The dielectric constant of the human body is approximately the same as that of water—about 80. Thus, when an intruder approaches the system, some of the electric flux will pass through his body, as shown in Fig. 6-1B. This has the effect of increasing the capacitance of the system. Since most of the body fluids are electrolytes, the human body is a rather lossy dielectric. Thus, the losses of the system also increase in the presence of an intruder.

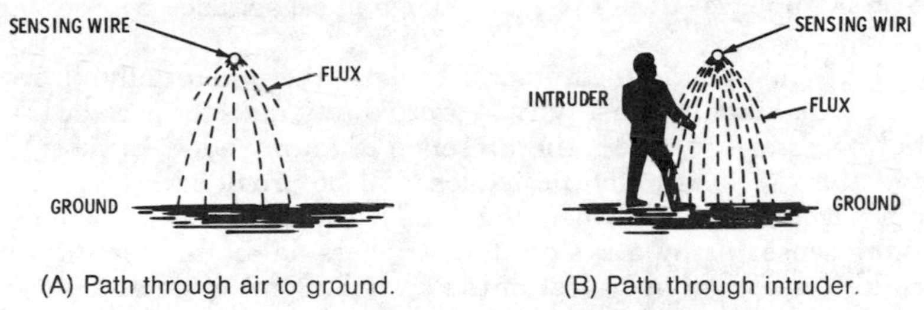

(A) Path through air to ground. (B) Path through intruder.

Fig. 6-1. Flux lines in a proximity detector.

The actual sensing device may be a straight wire, a wire that follows some path around the wall of a building, or a cabinet of almost any shape. An exact analysis of the situation involves electric field theory and can become very complicated. There are a few general principles, however, that can be used to improve the design of a system without involving much mathematical analysis.

The requirements for a sensing circuit for a proximity detector include the following:

1. The capacitance of the system, without an intruder present, should be as stable as practical.
2. The change in capacitance caused by the presence of an intruder must be great enough to trigger an alarm.
3. The system must not radiate enough energy to cause interference in the operation of other equipment.
4. The system should have minimum sensitivity to interference from other sources such as nearby radio stations.

In connection with the requirements of minimum radiation and susceptibility to radiation, it should be noted that the name "antenna" sometimes applied to the sensing wire is actually a misnomer. The sensing wire should not act at all like an antenna, which both radiates and receives electromagnetic energy.

Any wire longer than about 1/10 of a wavelength at the frequency of operation will *try* to act as an antenna. To minimize this effect, most intrusion detectors operate at frequencies below about 50 kHz. At this frequency, 1/10 wavelength is nearly 2000 feet, so a reasonably long sensing circuit will not radiate significant energy.

In order for the presence of an intruder to increase the capacitance of a system, some of the electric flux that flows between the "plates" of the capacitor must pass through the intruder. As a rule of thumb, the flux density varies inversely with the distance from the sensing wire. Thus, if the distance from the sensing wire to an intruder is cut in half, the change in capacitance will be four times as great.

This is not a serious limitation because there is usually no reason for making the sensitivity of a proximity detector greater than that necessary to detect the presence of an intruder. In most installations, just as much protection will be provided by a system that triggers an alarm when an intruder comes within a few inches of the sensor as by a system that triggers an alarm when an intruder comes within 20 feet of the system. Some systems are purposely made so insensitive that the intruder must actually touch the protected object to trigger the alarm.

Some idea of the capacitance of various arrangements of sensing circuits can be obtained from the formulas in Fig. 6-2. These formulas are approximate because the sensing circuit rarely has a neat geometry that will fit the equations closely. They will nevertheless give an idea of the order of capacitance that can be expected from various arrangements.

Many different circuits have been used to detect the change in capacitance of the sensing circuit of a proximity alarm. However, all of these arrangements operate on one of two principles. They

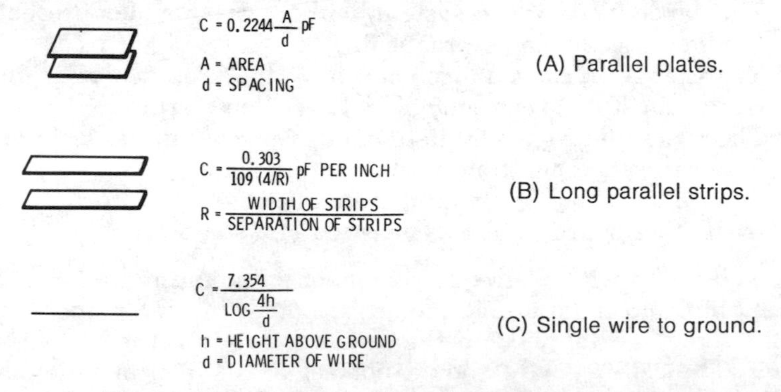

C = 0.2244 A/d pF
A = AREA
d = SPACING
(A) Parallel plates.
C = 0.303 / 109 (4/R) pF PER INCH
R = WIDTH OF STRIPS / SEPARATION OF STRIPS
(B) Long parallel strips.
C = 7.354 / LOG 4h/d
h = HEIGHT ABOVE GROUND
d = DIAMETER OF WIRE
(C) Single wire to ground.

Fig. 6-2. Capacitance equations for various arrangements.

either detect the change in capacitive reactance resulting from the capacitance change, or they detect the change in voltage across the capacitor that results from the capacitance change.

The capacitive reactance of any capacitor is given by the expression:

$$X_c = \frac{1}{2\pi fC}$$

where,

C is the capacitance in farads,
X_c is the capacitive reactance in ohms,
f is the frequency in hertz.

It isn't immediately obvious, but an inspection of this equation will show that the percentage change in capacitive reactance is the same as the percentage change in capacitance, regardless of what the capacitance is to start with or what frequency is used.

The most common way to detect the change in capacitive reactance is to connect the sensing circuit into the tank circuit of an oscillator. Then, any change in capacitance will result in a change in frequency, which is easy to detect.

Another way of detecting a change in capacitance is to apply a charge to the capacitor and detect the change in voltage that results from a change in capacitance. As shown in Fig. 6-3, the voltage across a capacitor is given by:

$$E = \frac{Q}{C}$$

where,

E is the voltage in volts,
Q is the charge in coulombs,
C is the capacitance in farads.

Therefore, if we increase the capacitance of the capacitor, the voltage across it will decrease, and vice versa.

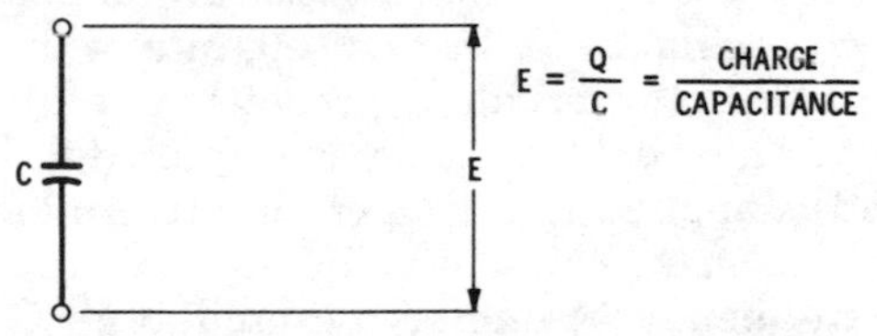

Fig. 6-3. Voltage, charge, and capacitance.

ADVERSE INFLUENCES

The capacitance between a wire and ground is influenced by any changes in the dielectric constant of the air between them.

The most common influence is a change in either the temperature or humidity of the air. A sensing circuit having a capacitance of 2000 pF may change as much as 50 pF at sunrise or sunset. This is a greater change than would be produced by an intruder in many systems.

To avoid false alarms resulting from changes in the quiescent capacitance of a system, ac coupling is usually incorporated at some point in the system. Such a system will not respond to very slow changes in capacitance but will respond to the faster change that occurs when an intruder suddenly approaches the sensing circuit.

Stray signals picked up by the sensing circuit are another influence. Although the sensing circuit is designed to be a rather poor antenna, it may still pick up a substantial signal from a nearby radio transmitter. Even though there may not be a radio station nearby, the increasing use of mobile radio means that a transmitter may suddenly show up at almost any location. To minimize this influence, a low-pass filter should be included in the system at the point where the sensing wire is connected to the remainder of the system.

Still another adverse influence is ordinary electrical noise of the type that causes interference with radio and television reception. Any electrical device, such as a motor or a neon sign, can cause enough electrical noise to produce false alarms. The alarm not only should have a low-pass filter at the point where the sensing wire connects to the remaining circuit, but should also have an interference filter at the point where the power line enters the unit.

BEAT-FREQUENCY PROXIMITY DETECTOR

Fig. 6-4 shows a block diagram of a proximity detector designed to compensate for the effects of temperature and humidity. This circuit uses two oscillators with two separate sensing wires. The oscillators are tuned to provide a predetermined beat, or difference, frequency. This beat frequency is amplified in a highly selective circuit and is applied to a trigger circuit. As long as the beat frequency remains at its original value, the alarm will not be initiated. If the frequency of one of the oscillators is changed, the beat frequency will no longer fall within the passband of the selective amplifier, and the input to the trigger circuit will decrease. This, in turn, will initiate the alarm.

The tuned circuits in the oscillators are designed so that when the third harmonic of one oscillator is heterodyned with the fourth harmonic of the other, equal changes in the capacitance of both

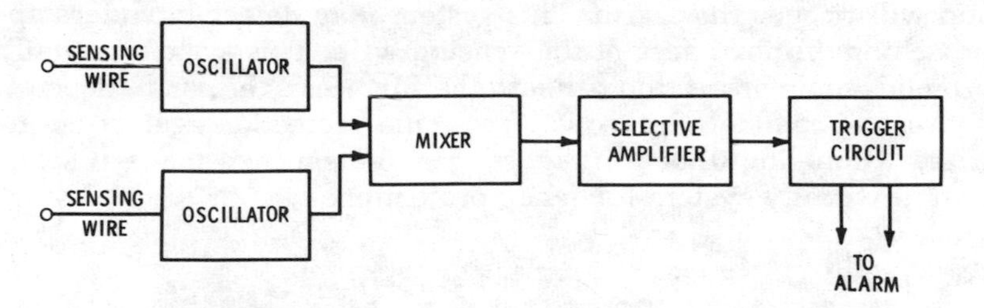

Fig. 6-4. Proximity detector with temperature and humidity compensation.

circuits will not change the beat frequency. If, on the other hand, the capacitance of one of the circuits is changed slightly, the beat frequency will change by a larger percentage. Suppose, for example, one oscillator is tuned to 100,000 Hz and the other is tuned to 74,875 Hz. The third harmonic of the first oscillator has a frequency of 300,000 Hz, and the fourth harmonic of the other oscillator has a frequency of 299,500 Hz. The beat frequency is then 500 Hz. If the frequency of the first oscillator changes to 100,010 Hz (0.01-percent change), the beat frequency will change to 530 Hz, a 6-percent change. Thus, the percentage of change in the beat frequency is 600 times as great as the percentage of change in the frequency of the 100,000-Hz oscillator.

This circuit may be used in an outdoor application such as that shown in Fig. 6-5, where the two sensing wires are run close to a fence to provide perimeter protection. Changes in temperature, humidity, and motion of the fence will produce equal capacitance changes so that the beat frequency will not change and false alarms will be prevented. On the other hand, if an intruder approaches, he will have more influence on the lower sensing wire

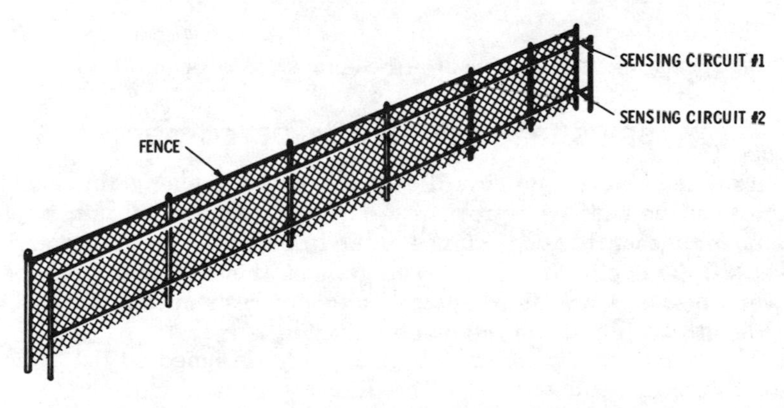

Fig. 6-5. Application of proximity detector along a fence.

and will trigger the alarm. This system may detect intruders approaching within 7 feet of the sensing wire. This system actually provides some protection beneath the surface of the earth, depending on soil conductivity, so it gives some protection against an intruder tunneling under the fence. Fig. 6-6 shows a modern solid-state proximity system for fence protection.

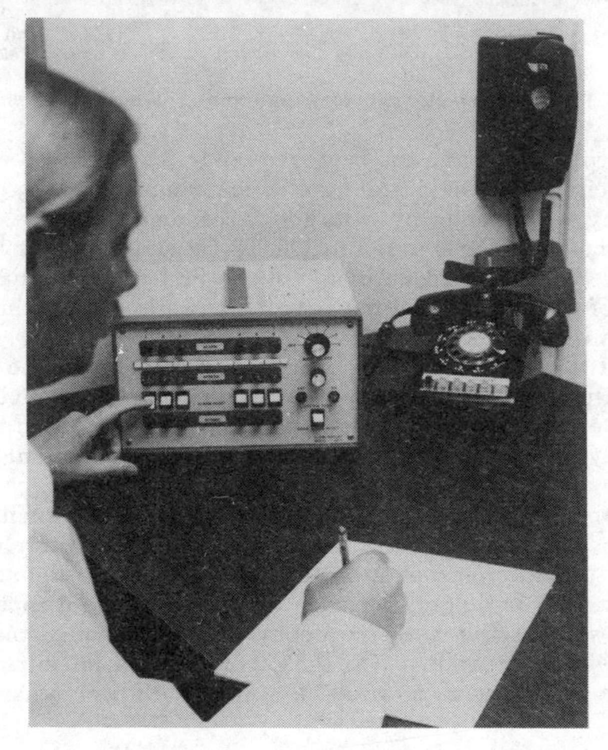

Courtesy GTE Sylvania, Inc.

Fig. 6-6. Console of a fence-protection detector.

BRIDGE-TYPE PROXIMITY DETECTORS

As noted before, any circuit that will detect a change in capacitance can be used as a proximity detector. One of the oldest systems for measuring capacitance is the impedance bridge shown in Fig. 6-7. This circuit is balanced—that is, there is no voltage between points A and B when the values of resistance and capacitance are such that the following equation is true:

$$\frac{C1}{C2} = \frac{R2}{R1}$$

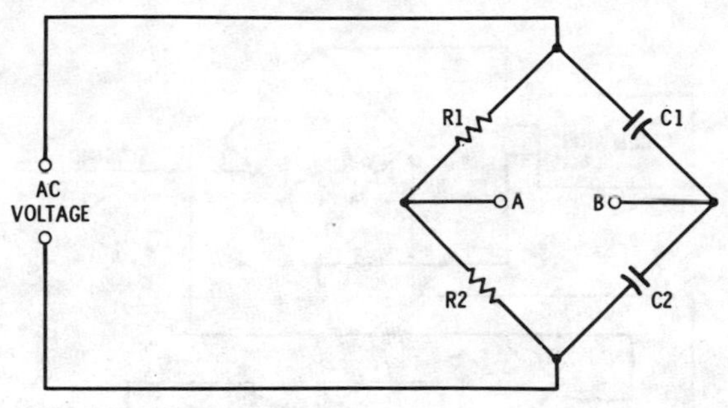

Fig. 6-7. Basic capacitance bridge.

When R1 = R2, the bridge is balanced when the two capacitances are equal. When the bridge is not balanced, an ac voltage appears between points A and B.

A practical version of this arrangement is shown in Fig. 6-8. Here, one of the capacitors is actually the sensing circuit. When the capacitance of the sensing circuit is equal to the capacitance of C1, the bridge is balanced and there is no voltage at the input of the trigger circuit. When an intruder approaches the sensing circuit, he causes its capacitance to change, unbalancing the bridge circuit. Then an ac voltage appears at the output of the bridge circuit. This voltage is rectified and fed to a trigger circuit which initiates an alarm. The circuit of Fig. 6-8 is commonly used in proximity alarms, but it has several limitations. In order to get good coverage, a fairly large signal is required from the oscillator. This may cause interference to radio or television receivers. The circuit is also subject to false alarms caused by stray signals picked up by the sensing circuit. If a strong signal is picked up by the sensing circuit, some of it may appear at the rectifier and trigger the alarm.

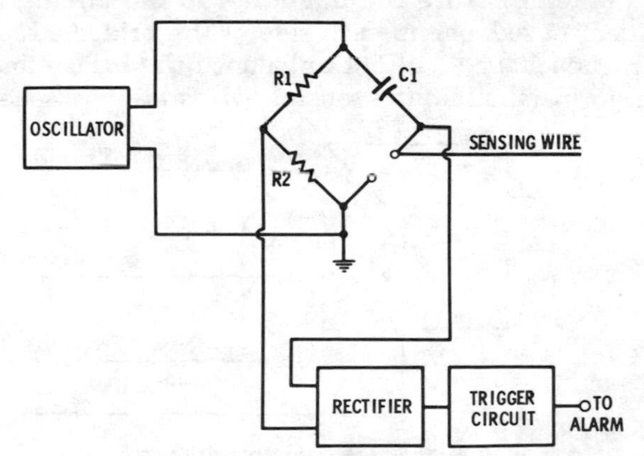

Fig. 6-8. Bridge-type proximity detector.

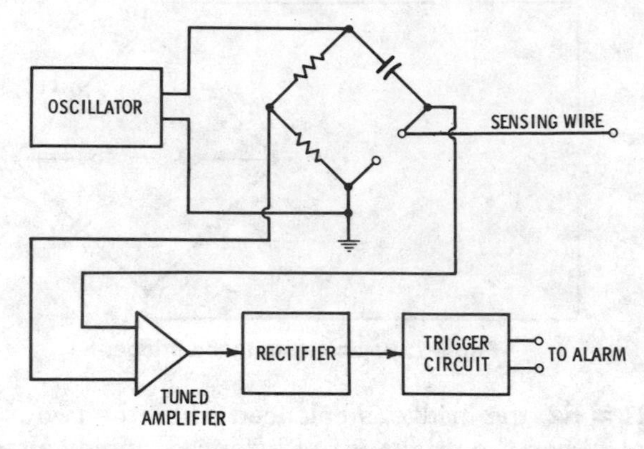

Fig. 6-9. An advanced bridge-type proximity detector.

The circuit shown in Fig. 6-9 is very similar to that of Fig. 6-8. The difference is that a tuned amplifier is used ahead of the rectifier and trigger circuit. This makes it possible to use a very low-level signal on the bridge circuit, reducing the probability that the system will cause interference. Since the amplifier ahead of the trigger circuit is sharply tuned, the circuit is also much less susceptible to interference. And since the components of the oscillator, tuned amplifier, and bridge circuit can be made highly stable, the chief cause of false alarms is changes in the sensing circuit itself. These can be minimized by careful installation and by not making the circuit any more sensitive than necessary.

Fig. 6-10 shows a bridge-type proximity detector that is designed particularly for outdoor applications, such as fence protection, where large changes in capacitance can be expected because of changes in temperature and humidity. In this circuit, two sensing circuits are used, one in each side of the bridge circuit. Equal changes in capacitance will not unbalance the bridge and initiate the alarm. In installation, the sensing wires are positioned so that

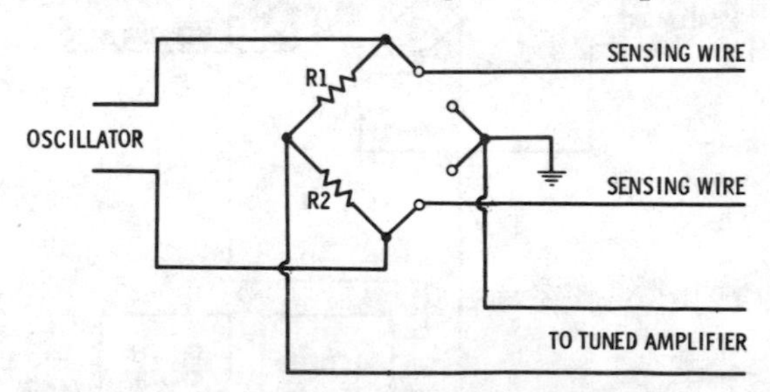

Fig. 6-10. A bridge-type detector with temperature and humidity compensation.

temperature and humidity changes will affect the two circuits in the same way, but an intruder will affect one circuit more than the other.

FET PROXIMITY DETECTORS

Earlier in this chapter, we saw that if we increased the value of the capacitance, the voltage across it would decrease, and vice versa. This suggests that we could make an intrusion alarm by merely measuring the voltage across a charged capacitor, which indeed we could, except that most voltage-measuring circuits would drain the charge from the capacitor. The insulated-gate FET has an extremely high input impedance, however, and can be used for this purpose.

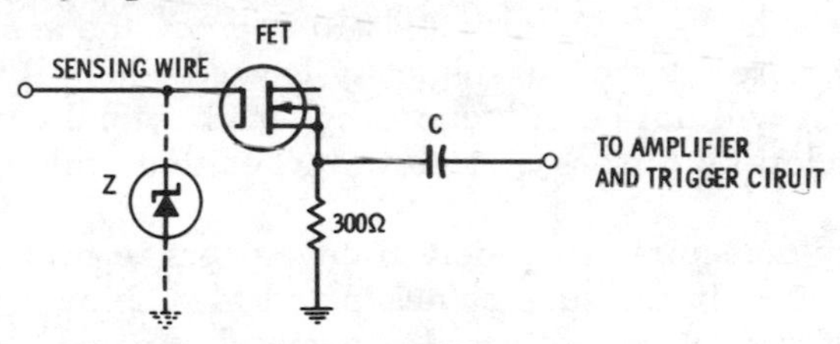

Fig. 6-11. Proximity detector using an FET.

Fig. 6-11 shows the circuit arrangement of an FET proximity detector. The sensing wire or antenna is connected to the gate of the FET. Since the gate is not directly connected to the rest of the semiconductor material, there is no way for the charge to escape. The amount of charge on the gate of the FET determines how much current will flow between the source and the drain, which are analogous to the cathode and plate of a vacuum tube, respectively. Since the sensing wire is insulated from the rest of the circuit, it will accumulate a small charge. When an intruder approaches, the capacitance between the sensing wire and ground will change. This, in turn, will change the amount of current flowing in the FET. The FET operates as a source follower, which is very similar to a cathode follower, and the current change causes a voltage change across the source resistor. This voltage is passed through a low-pass filter, amplified, rectified, and used to drive a trigger circuit which initiates an alarm.

The principal limitation of the FET proximity detector is its high sensitivity. It can be made extremely sensitive and is then subject to false alarms from temperature and humidity changes. Nevertheless, when properly installed and adjusted, it is a very effective proximity detector.

The proximity detector is well suited for both perimeter protection and protection of specific objects or small areas. In perimeter protection, the sensing wire is run around the protected area. In a typical example, it might actually be a part of a fence. The sensitivity can be adjusted so that the alarm will be triggered whenever anyone tries to enter the protected area. When properly adjusted, it will detect an intruder who tries to scale a fence without actually touching it.

Outdoor applications are subject to frequent false alarms unless great care is exercised in the application. The most common causes of false alarms are changes in temperature and humidity. Rain is particularly troublesome, but the effects of rain can be minimized by the use of high-grade insulators to support the sensing wire. The sensing wire must be strategically located where it will detect intruders but will not be influenced by small animals. It must not be located close to bushes or tree branches that will sway in the wind.

Proximity detectors are widely used indoors to provide protection for specific objects such as desks and file cabinets. Often, the objects that are to be protected are metal and can actually be made a part of the protective circuit. Fig. 6-12 shows a system in which the sensing wire is connected to a row of file cabinets that act as one plate of the capacitor in the sensing circuit. In most systems, the number of objects that can be protected by a single system is limited by the maximum capacitance that can be tolerated in the sensing circuit. Commercially available systems can protect up to 40 file cabinets connected together. In the arrangement of Fig. 6-12, one side of the system should be connected to a good earth ground or a cold-water pipe. In cases where a good

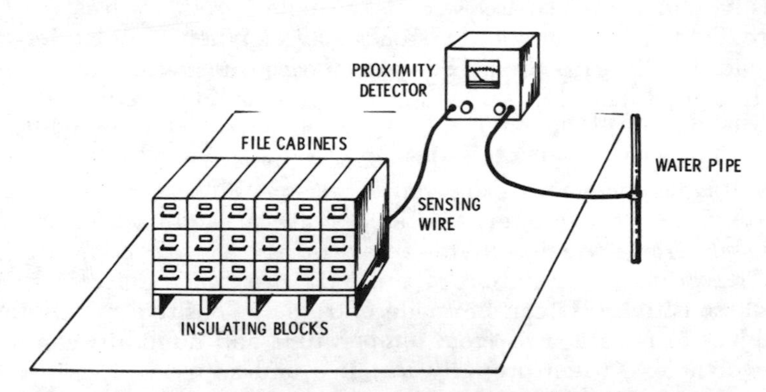

Fig. 6-12. Proximity detector used to protect file cabinets.

ground is not available, it is desirable to make a ground of screen wire on the floor under and around the protected objects, as shown in Fig. 6-13. The protected objects should not be any closer than 6 inches to the walls and should be mounted on insulating blocks. Loose wires such as lamp or telephone cords must be kept away from the protected objects because they are apt to be the cause of false alarms.

When it is impractical to locate the unit close to the area to be protected, the sensing circuit should be connected through coaxial cable. However, the capacitance of the cable will limit the number of objects that can be protected.

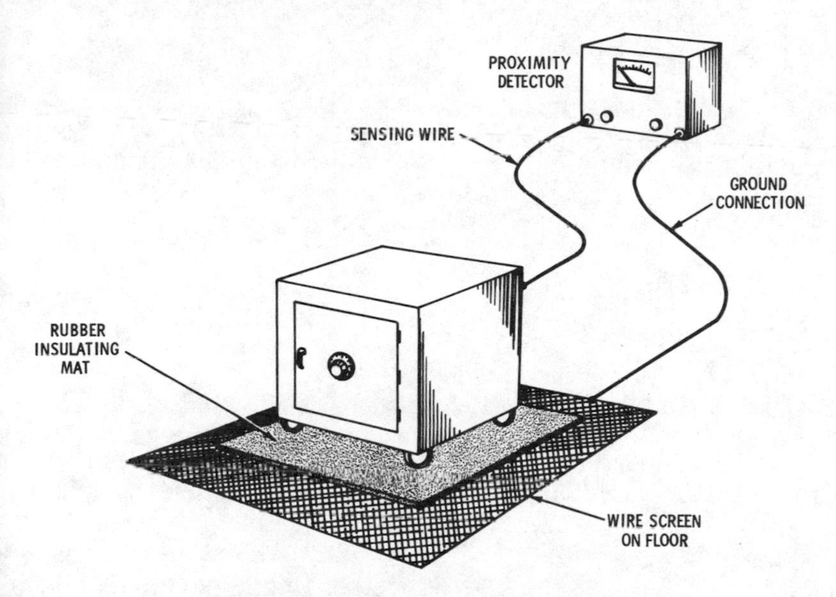

Fig. 6-13. Wire screen used as a ground.

ADVANTAGES AND LIMITATIONS

The proximity detector is very well suited to the protection of specific objects. One outstanding advantage is that the case of the unit itself may easily be made a part of the protected circuit so that the alarm will be initiated by any attempt to foil the system. The principal advantage of the proximity system is its versatility. It can be used to provide protection of almost anything, but will not be disturbed by normal activities a few feet away. Thus, safes, jewelry cases, and file cabinets can be protected so that an alarm will be initiated if anyone comes close to them, but normal business can be conducted close by.

The chief limitation of proximity systems is that they are very sensitive, and if the sensitivity in a particular application is made too high, there are apt to be frequent false alarms. Like other systems, they cannot be simply plugged in and expected to operate properly. The application must be planned to provide adequate protection with a minimum probability of false alarms.

CHAPTER 7

Audio and Visual Monitoring

One obvious way to make an area secure is to post guards in the area and at every entrance and exist so that they can watch and listen for any attempt to enter the area. In fact, this is done in many facilities where highly sensitive, classified work is being carried on. In most facilities, it is not economical to post guards so that all of the protected area is under their surveillance. It is both possible and practical, however, to use electronic systems to extend the listening and watching area of a security guard. Audio and video systems are widely used for this purpose. In many applications, audio and video systems are used not only for monitoring but also as an intrusion detector to trip an alarm whenever anyone enters the protected area.

APARTMENT BUILDING INTERCOM SYSTEM

Audio monitoring systems are widely used to protect apartment buildings from intruders. The common arrangement is a simple intercom system used in conjunction with an electrically operated door latch. With this system, shown in Fig. 7-1, the door of the apartment building is kept locked at all times. All tenants, and others who are authorized to enter, have keys. When a visitor not having a key desires to enter the building, he must press the doorbell button of one of the apartments. The tenant can then talk to the visitor over the intercom to establish his identity. When the tenant is satisfied that it is safe to admit the visitor, he can press a button that operates the electric door latch.

This simple system has many limitations. Probably its greatest disadvantage is that a single, careless tenant who will admit anyone who sounds his bell will jeopardize the security of the entire building. Another limitation is that a visitor can pose as a tenant who has misplaced his key and wait until another tenant opens the door and then enter the building at the same time. In most large apartment buildings, the tenants do not know each other personally.

In spite of its limitations, the apartment building intercom does provide some control over the number of people who enter and leave the building. If all of the tenants are security-minded, the system can be made very effective. It also acts as a deterrent to the burglar or vandal who does not wish to enter a place where he will have to call attention to his presence.

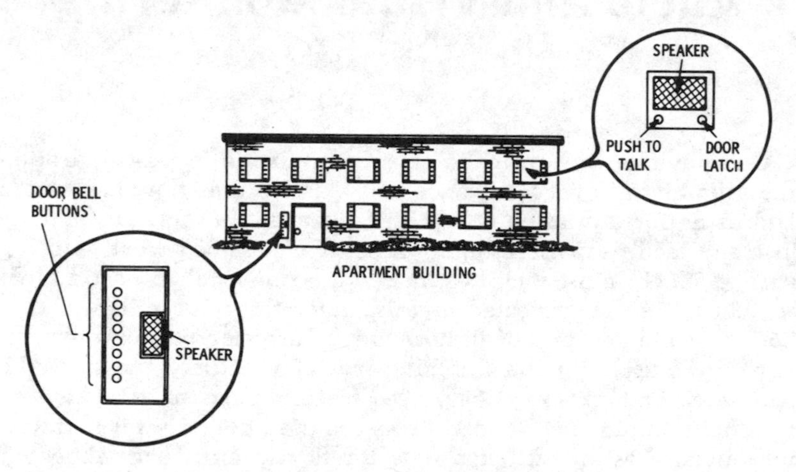

Fig. 7-1. Apartment building intercom.

GATE MONITORING SYSTEMS

A very similar system is frequently used to control access to industrial plants. With such an arrangement, a single guard can control access to several different entrances. Naturally, this system has many limitations. It is very difficult to positively identify a visitor by merely hearing his voice. Furthermore, a silent intruder may slip in with an authorized visitor. The effectiveness of the system is increased by proper procedures. For example, after a visitor has identified himself, he may be permitted to enter the plant but told to report to a receptionist or security guard immediately. In this way, if anyone enters but does not follow instructions, a security guard can immediately start searching for him.

AUDIO INTRUSION DETECTORS

In addition to being used for controlling access to doors or gates, audio intercom systems can be used to monitor an area for the presence of intruders. The regular intercom can be used for this purpose and, in fact, is often used in this way, but it has a disadvantage in that a guard must constantly listen for an intruder who is trying to be as quiet as possible. This is particularly difficult where one security guard must monitor several different areas. This disadvantage is overcome by adding a trigger circuit as shown in Fig, 7-2. In this system, a microphone (often a permanent-magnet speaker) located in the protected area is connected to an amplifier as in an ordinary intercom system, but instead of being connected to a monitoring speaker, the output of the amplifier is connected to a diode rectifier and a trigger circuit. Whenever the sound level in the protected area exceeds a preset level, the trigger circuit will operate, initiating an alarm.

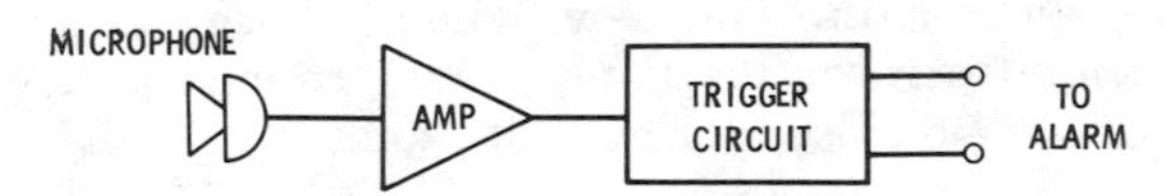

Fig. 7-2. Audio intrusion alarm.

The audio intrusion detector has many limitations. It is very susceptible to false alarms in locations where the background noise may vary over a wide range under normal conditions. Thus, it is not well suited for use in an area where machinery that operates all night turns on and off automatically, producing all sorts of different sounds. Neither is it suitable for protecting an area that has heavy drapes and thick carpeting. For example, a burglar might steal a typewriter from an office, making hardly any sound at all. The effectiveness of this system may be increased considerably by using doors, locks, and cabinets that cannot be broken without making a great deal of noise.

One outstanding advantage of the audio intrusion detector over other types is that it can be used to evaluate the situation causing the alarm. For example, if a guard at a central location is monitoring several areas, he can turn up the monitor speaker whenever an alarm is triggered. He can then listen carefully to the sounds in the protected area and determine whether or not there is an intruder in the area. If the alarm were triggered by some natural sound, such as a loose window rattling, he could probably recognize the situation. On the other hand, if he were to hear sounds that would not normally occur in the protected area, he could immediately send someone to investigate, or call the local police. This system has the advantage of detecting abnormal conditions other than intrusions. For example, the alarm would be triggered if a piece of machinery were to make an unusual noise because of a bearing failure.

A more sophisticated audio intrusion detector is shown in Fig. 7-3. This system has two additional features that will minimize false alarms as well as increase the versatility of the system. The first of these is a system of audio filters that can be adjusted to

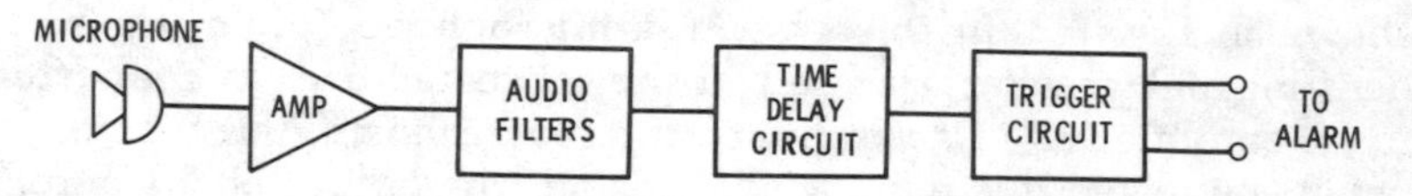

Fig. 7-3. Audio intrusion alarm with noise filter and time-delay circuit.

discriminate against the sounds that normally occur in the protected area. In addition, a time-delay circuit is provided that can be adjusted so that the alarm will not be triggered unless the abnormal sound persists for some predetermined period of time. This helps to avoid false alarms that would be caused by sounds such as thunder or a passing truck.

Sound Cancellation

Some audio intrusion systems use a circuit that will not trigger the alarm if the sound that is picked up originates outside of the building. This system, shown in Fig. 7-4, has two microphones—one in the protected area and another outside the building. Any loud sound originating outside the building, such as the sound of a passing truck, will be picked up by both microphones, but the circuit is arranged so that when the same sound is picked up by both

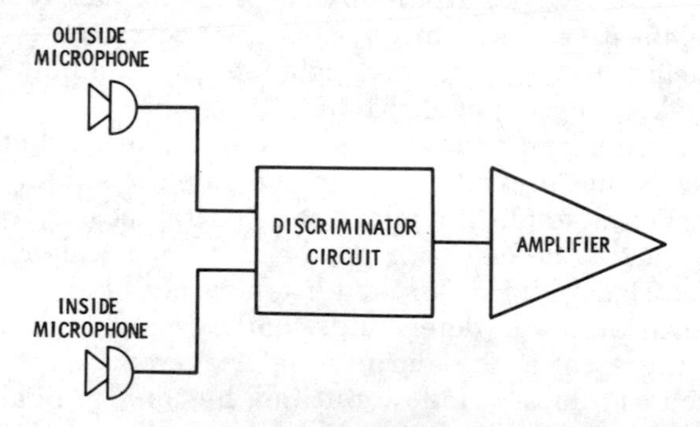

Fig. 7-4. A sound cancellation system to prevent triggering of alarm by loud exterior noises.

microphones, the alarm will not be triggered if the signal from the outside microphone is stronger. Thus, the circuit would not be triggered by a clap of thunder, but would be triggered by the sound of a safe being blasted inside the building.

By proper use of sound cancellation, filters, and time-delay circuits, the audio alarm can be adapted to many different applications. It must, however, be carefully tailored to the application. Before adjusting the system, the installer must know *all* sounds that will normally be present in the protected area at any time. If the installation is made during warm weather, care must be taken to ensure that false alarms will not be caused by the heating system later on. As with other systems, the final proof of effectiveness is a "walk through" test where every effort is made to foil the system.

Portable Audio Alarm

The audio alarm system can be made very compact. All that is required is a pm speaker and a simple circuit that may consist of a few transistors. The operating power may be supplied by a small battery. For these reasons, the audio detector is often used to provide temporary protection for an area that normally does not need protection. Suppose, for example, that a truck arrives at a plant too late to be unloaded and must remain overnight outside the building. An audio alarm can be mounted on the truck to provide protection. If the truck is properly secured, it cannot be opened without causing a lot of noise that will trigger the alarm.

VIBRATION INTRUSION DETECTORS

The vibration detector is practically identical to the audio intrusion detector except for the pickup device. Whereas an audio detector uses a microphone that will pick up sounds in the protected area, the vibration system uses a vibration detector that must be moved physically before it will produce a signal. The vibration detector is electrically about the same as a microphone, but it does not have a diaphragm.

Fig. 7-5 shows a sketch of a typical vibration detector. The mechanism is very similar to a permanent-magnet speaker. A mass suspended from a spring holds a coil in the strong field of a permanent magnet. Normally, the mass is at rest and there is no output from the device. Whenever the case is subject to vibration, the mass will not move because of its inertia and the magnet is moved with respect to the coil. This motion results in the magnetic field being cut by the coil, which in turn induces a voltage in the coil. A small transistor amplifier increases the strength of the signal to a level suitable for transmission to a remote amplifier and trigger circuit.

The circuitry used with vibration detectors is identical to that used with audio detectors, except that speakers are not used for monitoring. The sensitivity of a vibration detector can be adjusted over a wide range. It may be adjusted so that the alarm will be tripped by vibration ranging anywhere from a light touch to a hammer blow.

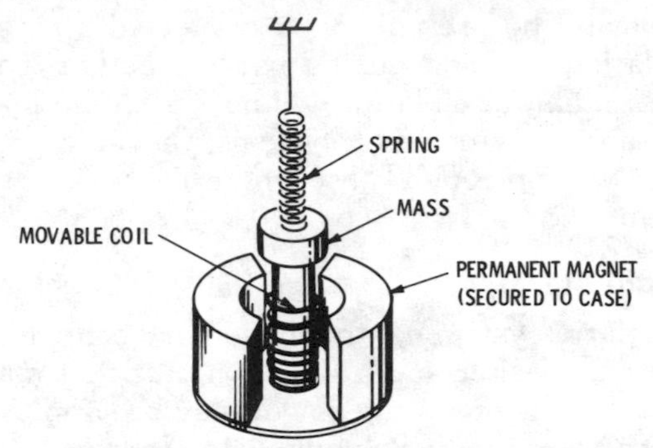

Fig. 7-5. A vibration detector.

All of the accessory circuits, such as filters and time delays that are used with audio intrusion detectors, may be used with vibration detectors. Vibration cancellation can also be used in the same way as sound cancellation to cancel out the effects of things that vibrate an entire building, such as the rumble of a passing train.

Application of Vibration Detectors

Although the principle and circuitry of the vibration intrusion detector are very similar to those of the audio detector, the application is completely different. Whereas the audio system is used primarily for area protection, the vibration detector is best suited for protection of specific objects such as file cabinets, safes, and similar objects. It is also well suited for use with other systems to protect against a burglar breaking through the walls of a facility not otherwise protected.

The effectiveness of the vibration detector depends upon proper application. It is often used to provide protection for a specific object in an area that might normally be occupied by people. For example, Fig. 7-6 shows a vibration detector mounted in a file cabinet to provide protection against anyone tampering with it. The file cabinet might be located in an office where cleaning people come in and out during the night when the office is not open for business. As long as they do not jar the cabinet excessively, the

Fig. 7-6. A vibration detector mounted in a file cabinet.

alarm will not be triggered. But if anyone tries to force the cabinet open, the alarm will be triggered. Proper adjustment of the sensitivity of the system is necessary to avoid false alarms.

VIDEO MONITORING SYSTEMS

Many of the limitations of audio monitoring systems can be overcome by the use of closed-circuit television systems. Although an intruder can remain perfectly still for a while without making any sound, he is still visible and can be seen on the television system. Monitoring of gates and doors is much more effective with television than with audio systems. Fig. 7-7 shows a typical camera installation in a museum. With several such cameras, a guard at a central location can monitor several parts of the museum.

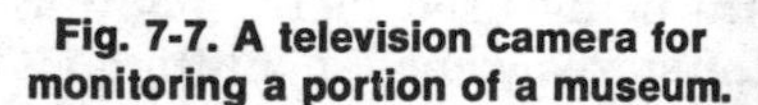

Fig. 7-7. A television camera for monitoring a portion of a museum.

Fig. 7-8 shows a television monitoring system where a security guard can monitor several warehouses.

When used in conjunction with other systems such as audio detectors, television systems are very effective for many applications, particularly where guards are on the premises. They are not suited for very remote locations because of the cost and complexity of transmitting television signals over long paths.

As an intrusion detector, the television system has the disadvantage that it must be watched continuously if an intruder is to be spotted when he passes the camera. This disadvantage can be overcome by using some other form of intrusion detector to detect the presence of an intruder, then evaluating the situation by watching the area on the television system.

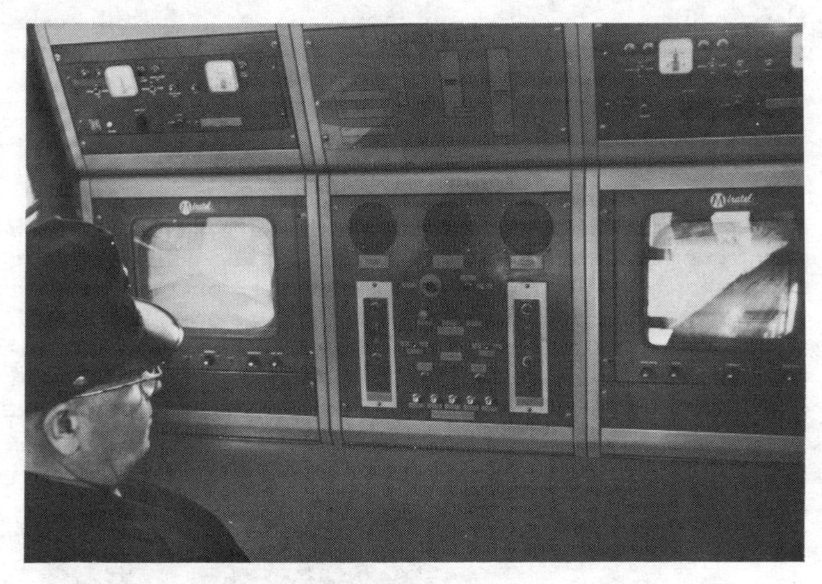

Fig. 7-8. A television monitoring system.

Television monitoring systems are being used increasingly for monitoring areas such as department stores where shoplifting and pilferage by employees are constant headaches. This arrangement permits a security officer to spot-check an area without anyone knowing he is being watched. This is a very effective deterrent. In fact, many decoy television cameras are in current use to discourage shoplifting. As was pointed out earlier, these systems will discourage amateurs but are an open invitation to the experienced shoplifter who knows that they are inoperative.

Television Motion Detectors

The requirement that a television monitor be watched continuously to detect an intruder can be overcome by the use of a video motion detector. The video motion detector operates in the following way. The video signal from the television camera at a particular period of time is sampled and stored in a memory circuit. At a later time when the camera is scanning the same area, the signal is sampled again and compared with the first sample. If nothing has moved in the area being scanned between the two samplings, the two samples will be the same. If, however, the picture has been changed by an intruder entering the area, the two samples will not be the same, and the alarm will be triggered. Fig. 7-9 shows the block diagram of such a system.

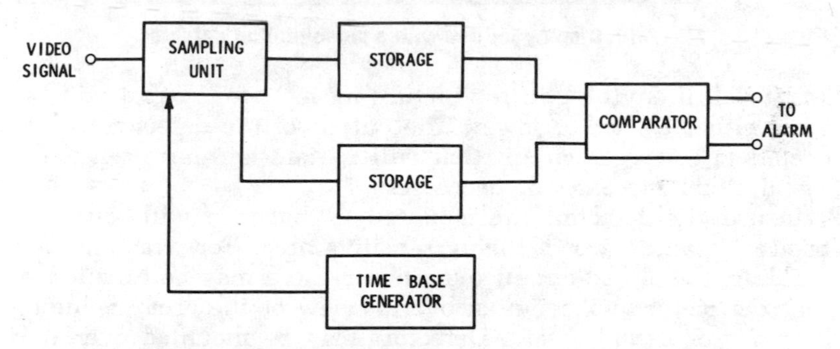

Fig. 7-9. Block diagram of a video motion detector.

The number of areas of the television screen that can be covered by a motion detector depends on the amount of storage capacity available. It is usually not considered practical to try to cover the entire screen. One way of showing the portion of the picture covered by a motion detector is to brighten that portion of the screen, as shown in Fig. 7-10. The sensitivity of the system is great enough to detect the motion of a person several feet from the camera.

A simpler form of video motion detector that is becoming popular is shown in Fig. 7-11. This system is not electrically connected to the television system. The detector (or detectors, since more than one can be used) is a photoelectric cell mounted within a container that is equipped with a suction cup so that it can be attached to the face of a television monitoring tube. The photocell integrates the light over the area covered by the suction cup. As long as the average light over this small area remains constant, the output of the detector remains constant. If the light changes

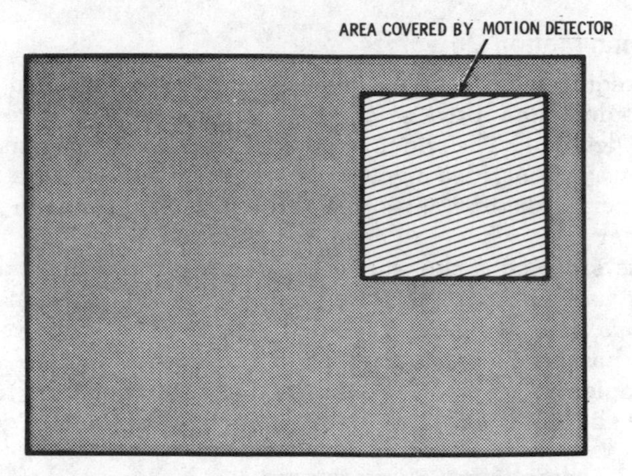

Fig. 7-10. Display for use with a video motion detector.

(as it will if anything in the picture moves), the output will become either higher or lower. The output of the detector is connected to a trigger circuit that will initiate an alarm when the signal either increases or decreases.

In use, the detectors are mounted so that they will cover entrances to an area or a highly sensitive area. For example, in a building, the closed-circuit television camera may be oriented so that the picture will give an overall view of the area, including both the door and a safe. Detectors may be mounted over that portion of the picture showing the door and the safe. Thus, if anyone enters the area or approaches the safe, an alarm will be sounded. A security guard will then have his attention directed to the screen. He can then watch the area to determine whether or not an emergency exists and take appropriate action.

The television system has the added advantage that it can be used to monitor all sorts of conditions other than security, such as the operation of equipment and the detection of other emergencies, such as a fire.

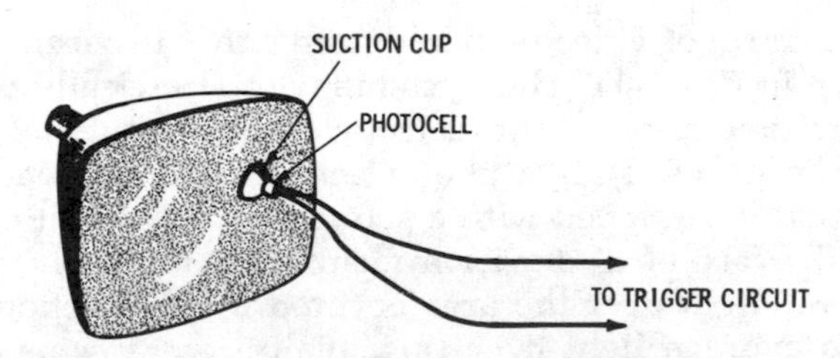

Fig. 7-11. A simple video motion detector.

Low-Light–Level Television

One of the traditional limitations of tv monitoring systems is that they are not very effective when the ambient light level is low. Recent developments in tv camera tubes, particularly silicon target and diode array tubes, have led to cameras that will provide satisfactory images when the light level is very low.

Courtesy Motorola Communications and Electronics

Fig. 7-12. A diode-array camera.

Fig. 7-12 shows a diode array camera that contains a low-noise preamplifier. This camera is particularly well suited to surveillance and monitoring.

A device that can be used in situations where there is little light is the infrared image intensifier shown in Fig. 7-13. This device is somewhat similar to a television camera tube and picture tube in the same package. When infrared energy strikes the photosensitive surface, electrons are released. The electrons are focused by a magnetic field onto a fluorescent screen at the opposite tube. This screen emits visible light. Thus, infrared energy, which is not normally visible, is converted into visible light.

When incorporated into a portable hand-held unit, the image intensifier system may be used to enable a security guard to inspect an otherwise dark area at night. The unit may also be used with a tv camera to permit viewing activities in dark areas or

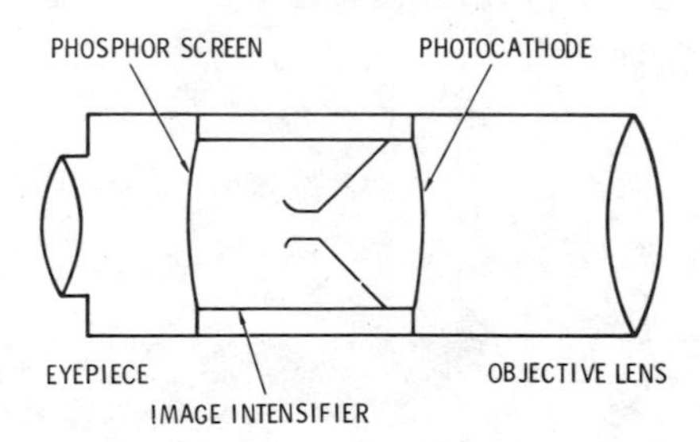

Fig. 7-13. Sectional drawing of an image intensifier.

(with a film camera) to photograph surreptitious activities for evidence or identification.

In areas where there is little illumination of any type, including infrared, an infrared spotlight may be used to provide the necessary illumination. The use of infrared, which is not visible, allows the intruder to think that he is operating in total darkness when, in fact, he is being observed.

CHAPTER **8**

Detection of Objects

There are many situations where security depends, not on detecting a person entering or leaving a facility, but on detecting the presence of an object that the person may be carrying. For example, it is perfectly normal for employees to pass through the gates of an industrial plant, but as they pass through it is desirable to know whether they are carrying stolen items of value concealed within their clothing. Similarly, it is normal for passengers to pass through the gates at an air terminal, but the security of the flight could be seriously affected if one of the passengers is actually a hijacker carrying a concealed weapon.

In most cases it is impractical, and it is often illegal, to search people entering or leaving a facility. For this reason, electronic systems are being used increasingly to detect concealed objects.

The increase of terrorism in recent years has led to the development of many different object-detection systems that will detect the presence of weapons or explosives.

The detection of shoplifting is another area in which object-detection systems are being widely used. A report issued by the National Retail Merchants Association claims that, out of a group of 500 shoppers who were monitored in a large store, 42 were found to be shoplifting. This works out to be a ratio of about 1 shoplifter for every 12 customers. The exact extent of shoplifting is not known because many shoplifters are not apprehended; but merchants agree that the problem is severe and is becoming worse.

There is a basic difference between the problem of detecting a concealed weapon or explosive and the problem of detecting hidden merchandise carried by a shoplifter. The terrorist carries a weapon that security people do not have access to until after the person has been apprehended. The shoplifter, on the other hand, steals merchandise that the security people have had an opportunity to tag. To some extent this makes the shoplifted merchandise easier to detect.

METAL LOCATORS

Many security systems that are used to detect concealed objects operate on the same principle as treasure locators that are widely used by hobbyists, and mine detectors used by military agencies to detect land mines.

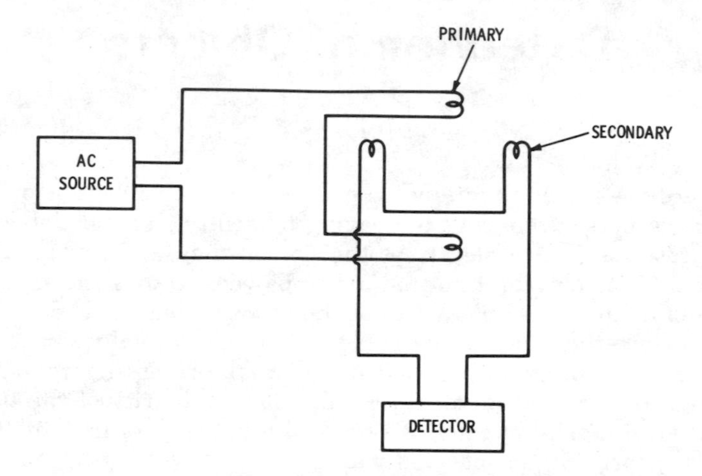

Fig. 8-1. Basic principle of a metal-object detector.

The basic principle of a metal locator is shown in Fig. 8-1. Here, an ac source is connected to one of two coils that are mounted at right angles to each other. The coils are constructed so that there is no magnetic coupling between them. That is, their mutual inductance is zero and no voltage is induced in the secondary coil. If a metallic object is placed in the field of the coil, as shown in Fig. 8-2, a voltage will be induced in it by the field of the primary coil. This causes eddy currents to flow, which in turn sets up a comparatively weak magnetic field. This weak field will induce a small voltage in the secondary coil. The voltage induced in the secondary coil is usually very small, so devices of this type are not very sensitive and large amounts of amplification are required.

There are three basic methods that can be used to detect the change in the circuit of Fig. 8-2 that is caused by the presence of the metallic object in the field of the coil.

1. Measurement of the impedance of the primary coil.
2. Measurement of the mutual inductance between the primary and secondary coils.
3. Measurement of the voltage induced in the secondary coil.

All three methods have been used in practical metal-object detector systems. The measurement of mutual inductance is the most popular method because changes in the geometry of the coils caused by temperature changes can be cancelled out easily and the resistive and reactive components of the induced signal can be easily separated. This method will discriminate between magnetic and nonmagnetic objects, which is often important.

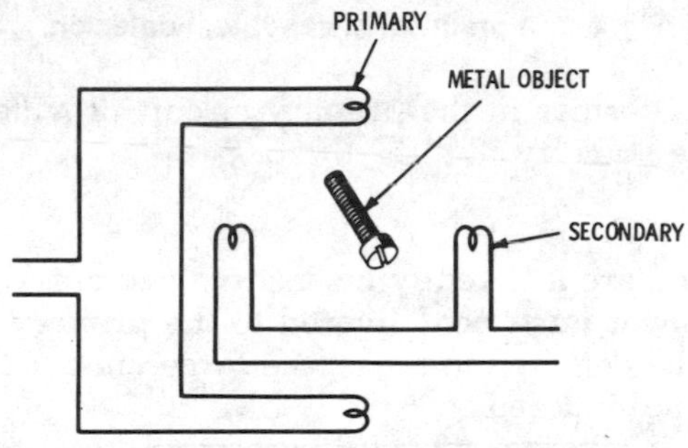

Fig. 8-2. Metal object placed in field of object detector.

The voltage induced in the metal object that is to be detected is proportional to the square of the frequency used. This would indicate that a higher-frequency system would be more sensitive, an indication borne out in fact. However, high-frequency systems are easily influenced by nonmetallic objects such as human bodies and moisture in the air. High-frequency systems are also influenced by stray fields, with the result that they often have shielding problems.

Although metal-locator systems have been made using frequencies all the way from about 15 hertz to the radio frequencies, the low audio frequencies are the most popular.

A PRACTICAL METAL LOCATOR

Fig. 8-3 shows the circuit of a practical metal locator that can be used for detecting the presence of concealed metal objects. The signal source in this case is actually a full-wave rectifier that develops a 120-hertz signal from the 60-hertz power line. This arrangement is more economical than using an oscillator, and has the added advantage that its frequency stability is good. A voltage-

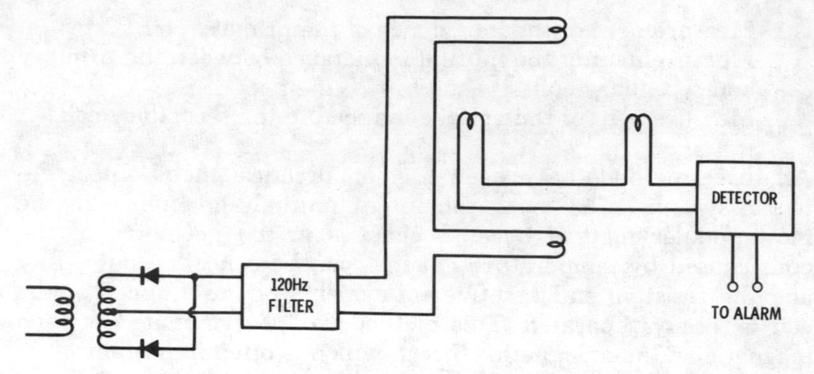

Fig. 8-3. A practical metal-object detector.

regulating transformer in the primary circuit provides the necessary amplitude stability.

Sensitivity

Metal locators are inherently insensitive devices. Usually about 100 watts of power must be delivered to the primary coil if metal objects are to be detected in any space large enough for a man to walk through unhindered.

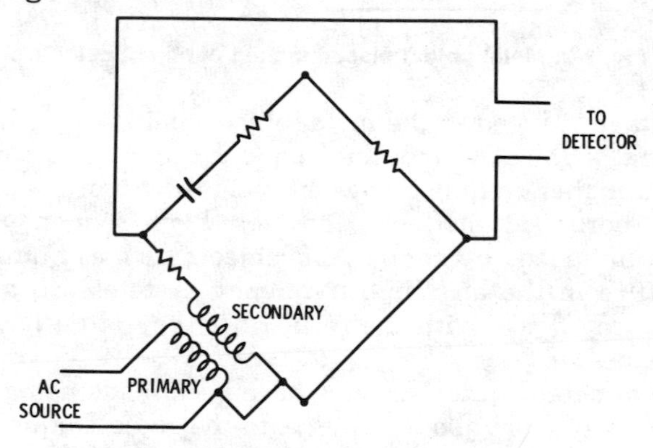

Fig. 8-4. Mutual-inductance bridge.

The detector is a mutual-inductance bridge, such as is shown in Fig. 8-4. The signal is amplified and detected by a phase detector that can be switched to distinguish between conducting objects and magnetic objects. The output of the phase detector is fed to a trigger circuit that initiates an alarm whenever a metal object is detected. The sensitivity of the circuit can be adjusted so that it will not trigger on small metallic objects, such as watches and cigarette lighters which anyone is apt to be carrying, but will initiate an alarm if a larger object such as a stolen tool or a concealed gun passes through the area.

The effectiveness of the metal detector as a security measure depends not so much on its technology as on the limitations in its use. In prisons, where the inmates have few rights, this device is very effective. A prisoner returning from a work area to his cell has no right to carry any metal objects at all. A metal detector that will be used with no restrictions can thus be made very sensitive so that it will detect the presence of even very small metal objects. If the alarm is initiated, the prisoner can be challenged and searched, if necessary.

In some national defense establishments, metal locators can be made almost as effective. Because of the much more important consideration of national security, employees give up many of their rights and will consent to being challenged and searched, if necessary. In most situations, however, people do have their rights, and, furthermore, in business it is essential to preserve their good will.

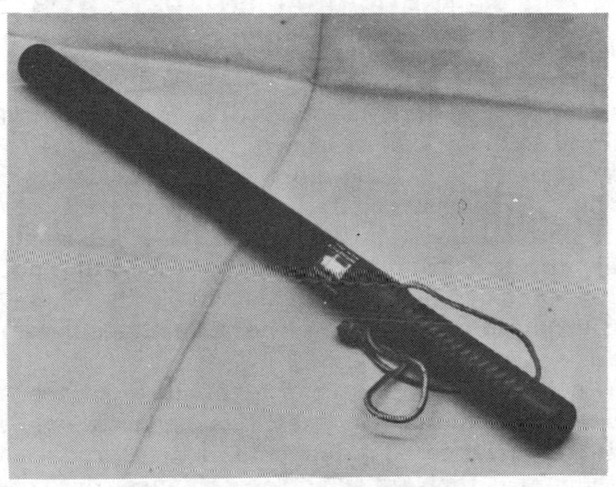

Fig. 8-5. A portable metal-object detector.

The metal detector will detect the presence of metal objects over a predetermined size. There is no difficulty here. The trouble lies in deciding what action should be taken after the alarm goes off. If the person passing the protected area is a customer, there is the question of incurring ill will by challenging him. If he is an employee, he may become sufficiently embarrassed to quit his job. Many industrial security officers feel that a metal locator is best used to obtain an idea of who is carrying concealed metal objects. When the alarm is initiated, no specific action is taken toward the employee, but the situation is noted and that employee is watched in the future.

At the present time, object detectors are not as widely used as intrusion detectors, but with the high rate of shoplifting from stores, and pilferage from industrial plants, we can expect to see many new developments in this area.

Portable Magnetic-Object Detector

Fig. 8-5 shows an object detector that is being used by police departments to search suspects when they are arrested. The device consists of a flux-gate magnetometer mounted in a policeman's nightstick. When the end of the nightstick comes close to magnetic material, it will cause an indication on the meter in the handle. The device has the obvious advantage that it permits its user to keep at a safe distance while searching a person for the presence of weapons.

THE MAGNETIC-GRADIENT DETECTOR

The magnetic-gradient detector was invented many years ago by the military for locating unexploded bombs. Recently it has found its way into security systems. The device consists essentially of three collinear coils as shown in Fig. 8-6. One of the coils is a transmitting coil, and the other two are receiving coils. The transmitting coil is connected through an impedance-matching network to an oscillator that produces a 100-kHz signal.

The magnetic field from the 100-kHz transmitting coil induces nearly equal voltages in the two receiving coils. These coils are connected in phase opposition, so the net voltage across the two coils is zero.

When a connecting object is brought close to the assembly, eddy currents are induced in it by the magnetic field from the transmitting coil. These eddy currents in turn produce magnetic fields of their own that either add to or cancel part of the field from the transmitting coil. As a result of this variation or gradient

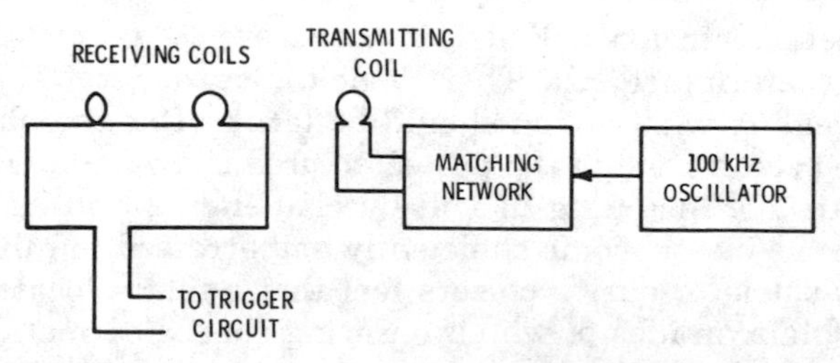

Fig. 8-6. Principle of a magnetic-gradient detector.

in the magnetic field, the voltages induced in the receiving coils are no longer equal, so they no longer cancel. The resulting voltage across the receiving coils is fed to a sensitive trigger circuit that sounds an alarm.

Magnetic-gradient detectors are used to detect weapons and explosives. They are also buried in the ground to detect the presence of passing vehicles.

DETECTING EXPLOSIVES

With the increase of terrorism, one of the most important applications of object detectors is the detection of hidden explosives. Terrorists can carry explosives aboard aircraft or into buildings. Detectors located at the entrances of these facilities will permit early detection of suspect objects. Bombs and explosives ranging from fairly large devices in packages to small letter bombs are sent through the mail. Object detectors can be arranged to screen packages and letters to detect such objects.

The principal consideration in the detection of explosives is that the object detector must not radiate or induce a great amount of energy into the suspected package. Whereas a firearm, such as a gun, will withstand a rather large magnetic or electric field without firing, many bombs and explosives are critical in this respect, and there is a possibility that the device that is intended to detect them will actually detonate them.

Systems designed to detect explosives should induce no more energy into the area being searched than is absolutely necessary. Thus, the detection circuits must be very sensitive. Since sensitive circuits are subject to triggering by external influences, the design is usually a compromise.

DETECTING TAGGED OBJECTS

Merchandise in a store can be tagged in some way to make its presence easier to detect. Fig. 8-7 shows the principle of a system designed to detect tagged objects. The detection circuit consists of an oscillator that is swept in frequency across a considerable portion of the spectrum—say over a range of 2 MHz in the high-frequency portion of the spectrum. A circuit is provided to detect the amount that the oscillator is loaded by an external circuit. A sensing coil connected to the oscillator provides an induction field that covers a large area.

The merchandise that is to be protected carries a tag containing a small resonant circuit. As the frequency of the oscillator sweeps through the resonant frequency of the tag, the oscillator will be

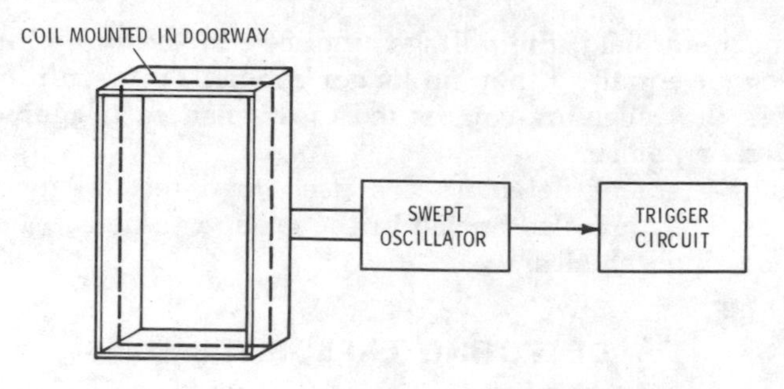

Fig. 8-7. A system for detecting tagged items.

loaded and a signal will be delivered to the trigger circuit which initiates an alarm. The entire arrangement works in much the same way as the old fashioned grid dipper.

These tags are usually fastened to the object to be protected by means of a strong nylon link which is difficult to break. When an item is purchased, the salesperson cuts the tag free when the item is paid for. If a shoplifter carries a tagged item through the door of the store, an alarm will sound. Similar tags are fitted into the bindings of books to prevent the book from being taken out of the store or library without being properly checked out.

The system as described has some disadvantages. A shoplifter may remove the tags from several items of clothing, place the clothing under his or her street clothes, and then leave the store without buying anything. This type of action can be frustrated to some extent by locating detectors in the fitting rooms. If a tag is found in the fitting room after a customer has finished trying on a garment, it is obvious that the customer has removed the tag from the garment. The detector in the fitting room will alert the store's security personnel.

Another disadvantage of the tagging system is that it can be thwarted if a cashier in a store cooperates with a shoplifter. This action can be checked by arranging a detector so that the cashier must insert a tag in an opening each time that an item is rung up on a cash register.

A variation of the tagging system is shown in Fig. 8-8. Here the detection system consists of a microwave transmitter with a receiver tuned to the second harmonic of the transmitter frequency. The tag is a small resonant device containing a semiconductor. When the tag is radiated by the transmitter signal, the semiconductor diode generates harmonic energy that is picked up by the receiver. The receiver then triggers an alarm.

When any radiation device is used in the vicinity of human beings, its effect on persons wearing a cardiac pacemaker must be

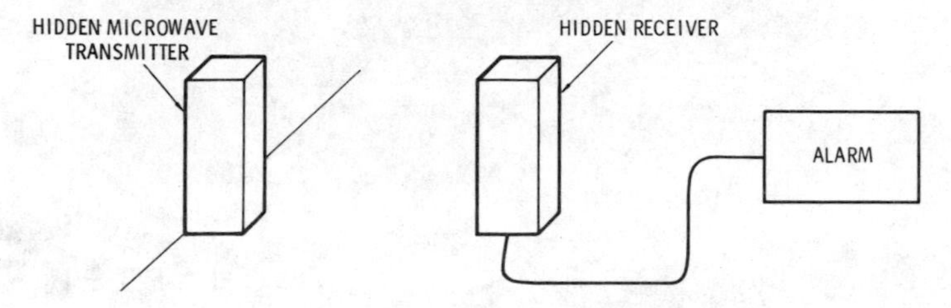

Fig. 8-8. A microwave tag detector.

considered. The Food and Drug Administration has investigated this matter, and no system should be used unless the manufacturer can guarantee that it will not interfere with the operation of pacemakers.

Of course, any device that radiates rf energy must comply with the FCC rules.

CHAPTER 9

Alarm and Signaling Systems

The alarm systems described in this chapter range from a simple bell or motor-driven siren to a complex system that will automatically dial the police department. Since almost all the intrusion detectors described in the previous chapters produce a simple alarm-initiating signal, almost any intrusion detector can be used with any of the alarms described here. It is quite common for several different types of intrusion detectors to be connected to a single alarm system.

LOCAL ALARMS

A local alarm is just what the name implies—an alarm located on the premises to be protected. Local alarms in general are not as effective as the more sophisticated systems to be described later, but they definitely have their place in security systems. The local alarm has two functions:

1. It must attract attention—the attention of either the police, the proprietor, or some civic-minded citizen who will call the police.
2. It should frighten or at least unnerve the intruder so that he will either abandon his plans or will take so much time that he is caught.

Typical Circuits

Fig. 9-1 shows a typical circuit used for operating a local alarm. Normally, the alarm contacts in the intrusion detector are closed.

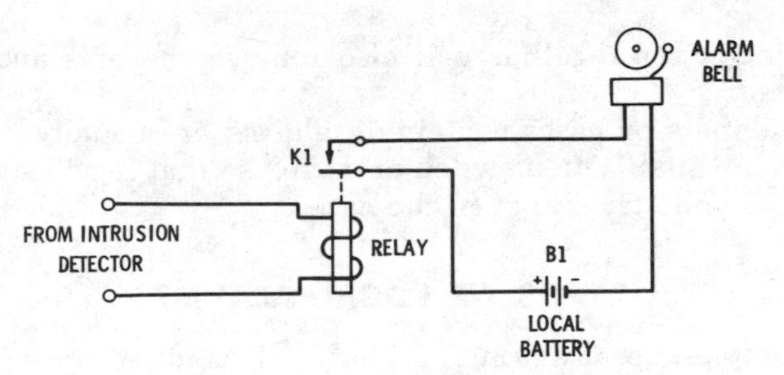

Fig. 9-1. A simple alarm circuit.

The current through the connecting wires will hold relay K1 closed and the contacts open. When the intrusion detector is actuated, the alarm contacts in the detector circuit open. This de-energizes relay K1, closing the circuit to the alarm. The connecting wires must be well protected with conduit and run inside the protected area. This system could be foiled if the two connecting wires at the bell were shorted together or if one of the wires were cut. In spite of this limitation, this arrangement is often used in installations where the connecting wires cannot be reached without tripping whatever intrusion detector is used.

In installations where the possible gain from burglary is very great, a burglar might think it worth the time and risk to drill through a wall to reach the connecting wires leading to the bell. In a case like this, additional protection can be obtained by using the circuit of Fig. 9-2. Here, a bias-voltage source is included in the intrusion detector. This system is almost identical to the electromechanical system described in Chapter 3. In this system, cutting the connecting wires will remove the bias on transistor Q1, allowing it to conduct and thus setting off the alarm. Shorting the

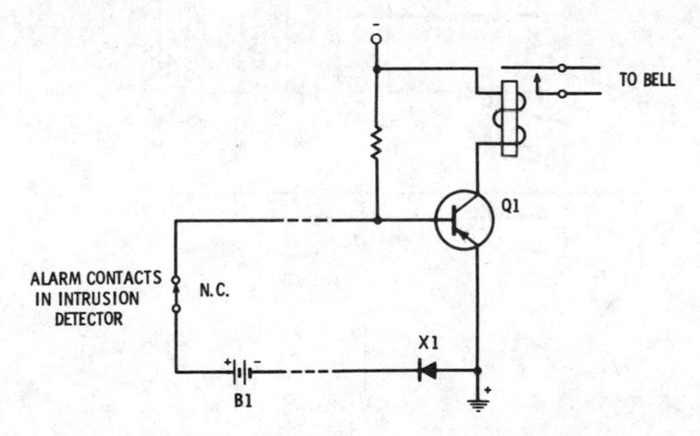

Fig. 9-2. An alarm circuit that trips on either open or short circuit.

connecting wires together will also remove the bias and set off the alarm.

The cabinets or cases containing alarms are usually equipped with a small snap-action switch arranged so that it will set off the alarm if anyone tries to open the case.

TYPES OF LOCAL ALARMS

The simplest local alarm is a loud bell mounted on an outside wall of the premises. Usually a separate battery is supplied for operating the bell so that it cannot be foiled by cutting the power lines.

Another type of local alarm that is gaining popularity is the electronic siren. This circuit produces a wailing sound and is often used as a siren on police cars. The siren has a pronounced psychological effect simply because it sounds like a police car. It would make an already nervous intruder even more nervous so that he would be less likely to complete his planned theft. Another factor is that since the alarm is a siren, it will tend to mask the sound of approaching police cars, increasing the chance that the intruder will be caught.

Fig. 9-3 shows the circuit of an electronic siren. In this circuit, unijunction transistor Q1 is a sawtooth generator that generates

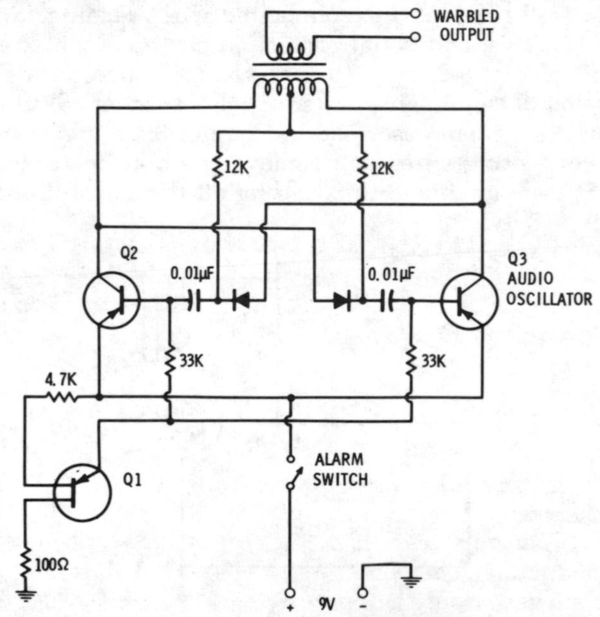

Fig. 9-3. An electronic siren circuit.

a voltage having the waveform shown in Fig. 9-4A. This waveform changes the frequency of an audio oscillator consisting of transistors Q2 and Q3 and their associated circuitry. The result is that the audio-frequency output periodically increases in frequency, then suddenly jumps back to its original frequency. This produces the familiar wailing sound of the electronic sirens used on many police cars. The output from the circuit shown is less than 1 watt. This is not loud enough to attract any attention, but the circuit can be followed by an audio power amplifier of the type used in public-address systems to provide an adequate sound level.

High-powered electronic sirens are among the most effective alarms. They can be made almost invulnerable to foiling if they are mounted in cases that are protected by interlock switches. Of course, there are more components in an electronic siren than in a simple bell, and consequently more possible things to fail. However, the high reliability of solid-state circuits will minimize the possibility of circuit failure. A test switch should be provided so that the user can be sure that the circuits are working properly before turning on the system for the night.

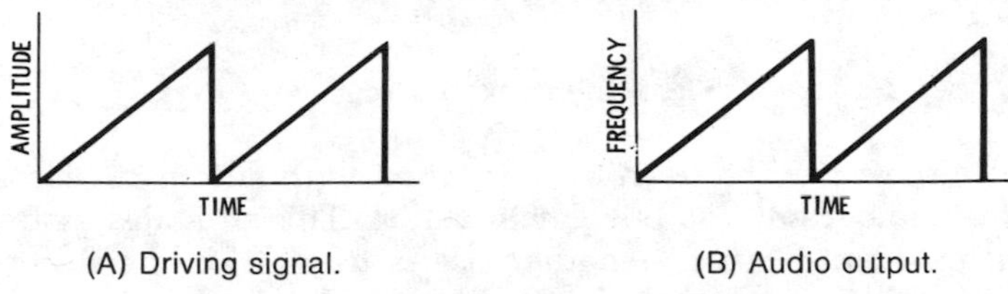

(A) Driving signal. (B) Audio output.

Fig. 9-4. Waveforms of an electronic siren.

CONTROL OF LIGHTS

A local alarm is frequently connected to the lights in the protected area. Then, when an intruder trips the alarm, the protected area is flooded with light. This has the double advantage of frightening the burglar and enabling the police to see what is going on when they respond to the alarm.

A variation of this light control is the addition of flashing bright lights that are set off in the protected area by the intrusion detector. These lights have a pronounced psychological effect on the intruder when he trips the alarm—blinding lights flash in his face while he is attempting to perform the robbery.

Fig. 9-5 shows a circuit of the type normally used in flash photography. Under quiescent conditions, the SCR and the flashtube are not conducting. The 2-μF capacitor is charged to the full value of the dc power-supply voltage. When a positive-going pulse is applied to the input circuit, it triggers the SCR into conduction.

The sudden surge of current through the SCR induces a high voltage in the secondary of the transformer. The secondary of the transformer is connected to a coil consisting of about six turns wrapped around the flashtube. The high voltage in this coil ionizes the gas in the flashtube, causing it to fire. All of the charge in the 2-μF capacitor then discharges through the flashtube, causing a bright flash.

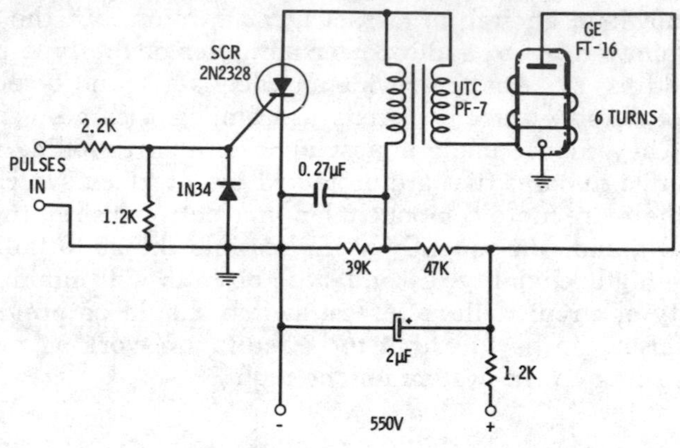

Fig. 9-5. Photoflash circuit.

This circuit may be used in connection with automatic cameras to get a photograph of a burglar in action. This provides evidence for future identification. Another use is to position the flashtube so that it will flash directly in the burglar's face when he touches a safe or some other specific object to be protected. In this case, the circuit can be driven by a low-frequency pulse generator such as that shown in Fig. 9-6. The flash rate may be adjusted by chang-

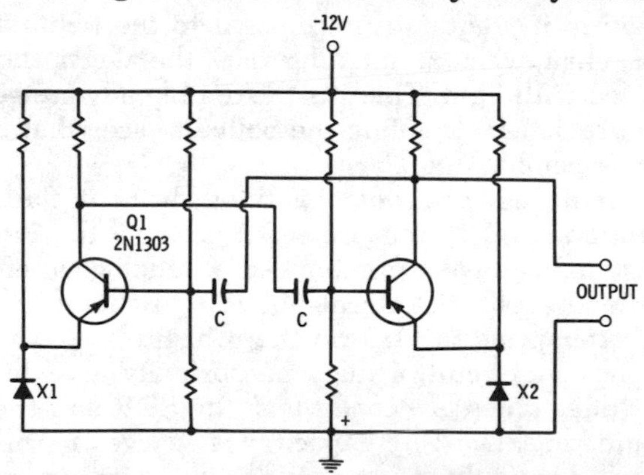

Fig. 9-6. Pulse generator used to drive a photoflash unit.

ing the value of capacitor C in Fig. 9-6. A flash rate of two to five flashes per second is not only blinding, but very disconcerting.

REMOTE MONITORING

In large facilities such as industrial plants, government installations, and schools, it is usually necessary to monitor the security status of the entire facility at one central location. This might be either a central point on the same premises, or the headquarters of a private security company several miles distant. In any case, the following indications are required at the monitoring point:

1. Normal or Secure. This is an indication that the system is turned on and is operating properly.
2. Alarm. This indication shows that an intruder has entered the protected area.
3. Trouble. This indication shows that there is trouble with the system, caused either by component failure or an attempt to foil the system.
4. Access. This indicates that the system has been disabled to allow authorized people to enter the protected area.

In addition to the foregoing indications, many installations provide an automatic printout of the time and date that each of the indicated functions changes. Some installations also have controlled door locks, intercoms, and closed-circuit television systems connected between the central monitoring point and each of the protected areas.

When the monitoring point is not too far from the protected areas, there are usually separate wires run from each intrusion alarm. When large distances are involved, it is usually more economical to use a coded system that transmits information from all detectors through a single pair of wires.

Security monitoring consoles vary in complexity with the security needs and size of the installation where they are used. The monitoring point may consist of a simple indicator in a guard's shack, or it may be an elaborate console in which a security officer can trace the progress of an intruder through a facility while directing police to the point where he will be intercepted.

Remote Monitoring Circuit

Fig. 9-7 shows a circuit to remotely monitor the security status of an area. In normal operation, the access switch is in the SECURE position and the alarm contacts are closed. Under this condition, there is a 6-volt drop across zener diode X1 in the protected area, and the voltage across the voltmeter is 6 volts. The voltmeter is

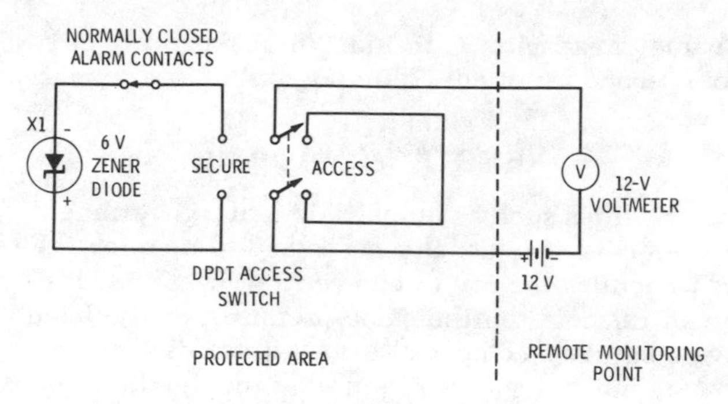

Fig. 9-7. A simple remote monitoring circuit.

shown in Fig. 9-8. The scale is marked so that at midscale (6 volts) it will indicate a secure condition. That is, the intrusion alarm has not been tripped. When the alarm contacts are opened by the presence of an intruder, the circuit will be opened and the voltmeter will show a zero-scale indication that is labeled "alarm."

When the access switch is thrown to the ACCESS position to allow authorized persons to enter the protected area, the switch is shorted, so there will be no voltage drop across it. The voltmeter will then show full-scale deflection, indicating an access condition. The guard monitoring the console will then know that the area in question is no longer protected.

One disadvantage of the arrangement as described so far is that a short-circuit across the lines from the protected area will cause the same indication as throwing the access switch to the ACCESS position. This disadvantage is overcome by adding a test switch and a second zener diode connected as shown in Fig. 9-9A. When the test switch is in the NORMAL position, the circuit is identical to that of Fig. 9-7. The addition of X2 across the ACCESS position will not affect the operation, since it is backward and there will be little voltage drop across it. This switch merely reverses the polarity of the battery and the voltmeter when in the TEST position. When

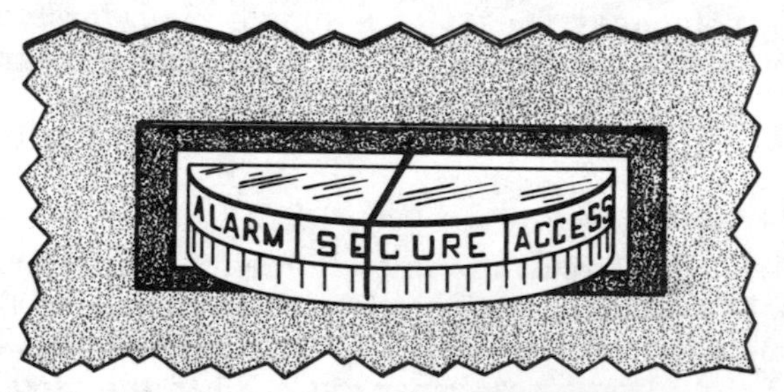

Fig. 9-8. Indicator scale used with security system.

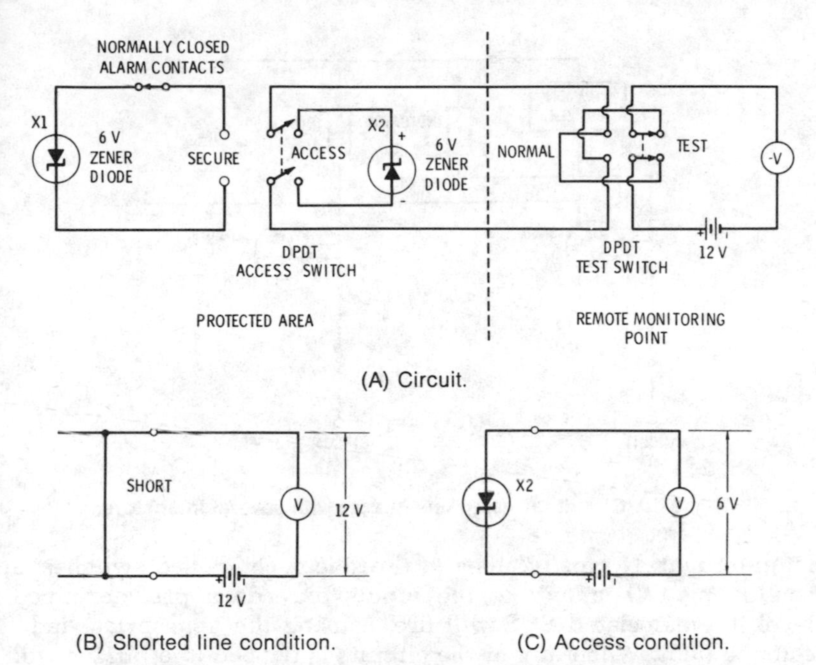

(A) Circuit.

(B) Shorted line condition. (C) Access condition.

Fig. 9-9. Remote monitoring circuit with provision for testing between "access" condition and shorted line.

the polarity is reversed (switch in TEST position), a short circuit across the line will form the circuit shown in Fig. 9-9B. The voltmeter will still read full scale in the SECURE position. If, however, the access switch is in the ACCESS position, the circuit will be as shown in Fig. 9-9C. Zener diode X2 will then be connected in the circuit in such a way as to reduce the voltage across the meter to 6 volts, and a "secure" reading will be obtained. Thus, by using the test switch, the guard can distinguish between an actual access condition and a short on the line.

Many variations of this basic remote-monitoring circuit are in common use. One of the most common is the use of voltage-sensitive circuits and indicator lamps instead of meter-type indications.

Local Buzzer or Bell

In addition to the meter or indicator-light monitor, it is desirable to have some sort of audible annunciator such as a buzzer or a bell that will alert the guard to the fact that the status of one of the indicators has changed. It is not necessary to use a separate buzzer for each indicator. Fig. 9-10 shows a simple circuit using silicon-controlled switches that will operate a buzzer whenever any of the channels is activated, but will cause only the proper

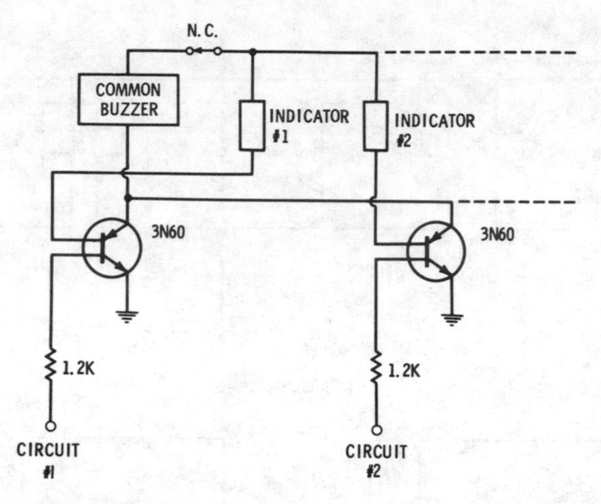

Fig. 9-10. Circuit for using one buzzer with several indicators.

lamp to light. Normally, none of the silicon-controlled switches is conducting. When any of the inputs exceeds a predetermined level, the associated SCS will fire, causing the appropriate indicator to light. When any of the circuits is tripped, the buzzer will sound. A switch is provided to permit the guard to silence the buzzer after it has attracted his attention.

Coded Signaling Systems

In many applications, the security monitoring point is located several miles from the protected areas. This is particularly true when a private security organization provides the protection. In such cases, it is usually more economical to use a single pair of wires from several different protected areas to the monitoring point. This is accomplished by using a code system that will identify each protected area and its security status.

A simple system that will accomplish this goal is shown in Fig. 9-11. Here, at each protected area or at some central point near several protected areas, a coder is located. This consists of a group of oscillators that generate different frequency signals correspond-

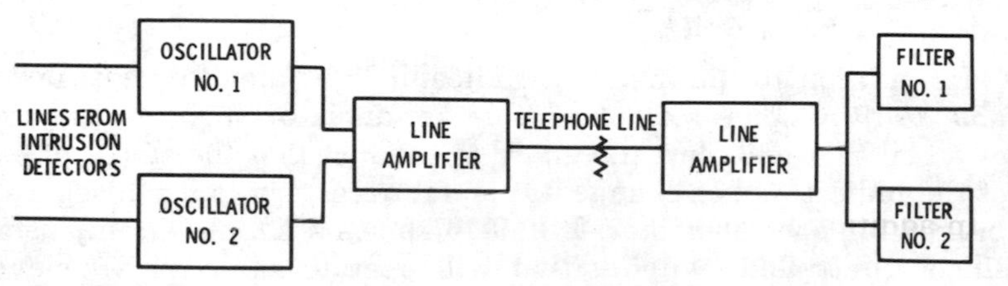

Fig. 9-11. System for transmitting several signals over one telephone line.

ing to the security status of each protected area. These signals are then combined and transmitted to the monitoring point over a single pair of wires. At the monitoring point, the signal is separated into its various components by filters. The output of each filter is then fed to the corresponding indicator.

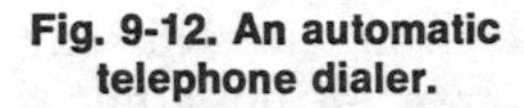

Fig. 9-12. An automatic telephone dialer.

AUTOMATIC TELEPHONE DIALING

One of the latest and most sophisticated additions to intrusion-alarm systems is a device that will automatically dial the local police department or security organization in the event of an intrusion. To some extent, this arrangement gives the small facility some of the advantages formerly enjoyed only by firms that could afford their own security forces.

An automatic dialing system is shown in Fig. 9-12. To understand how this system works, we must know something about the workings of a dial telephone system. A diagram of a telephone dial is shown in Fig. 9-13. Switch A connects the phone to the line. It operates whenever the phone is lifted from the cradle. Switch B opens the circuit to the receiver and shorts out the transmitter. This switch operates whenever the dial is moved from its normal talking position, and remains in this position until the dial gets back to its normal position. Switch C is called the *impulse switch*. By opening and closing, it generates the dial pulses. Note that the dial pulses are actually circuit interruptions. For example, dialing the number 5 will open the line five times. The slowest dialing rate in use today is about 12 interruptions per second.

TOUCH-TONE DIALING

In addition to conventional telephone dialing, "touch-tone" dialing is becoming increasingly popular throughout the country, and this principle is also used in automatic dialers.

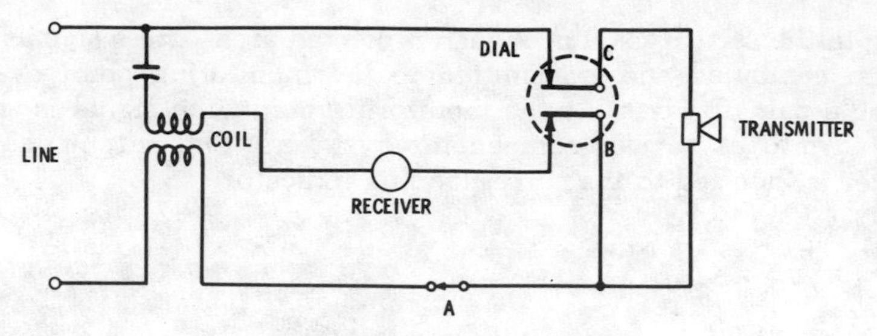

Fig. 9-13. Telephone dial circuit.

The touch-tone system operates by transmitting various tones in sequence. In order to minimize interference from other signals, two separate tones are combined to represent each number. The frequencies used to represent each number are listed in Table 9-1.

The following specifications should be met by any device that is used to dial automatically using this system:

1. The frequency of the various tones must be held to within one percent.
2. The amplitude of the dialing signal must be held within the limits specified by the telephone company.
3. Each double tone should have a duration of at least 40 milliseconds.
4. The maximum dialing rate should not be faster than 12 tone bursts per second.

Sequence of Events

The sequence of events involved in dialing a telephone call is as follows:

1. The telephone is connected to the line.
2. A short time later, the familiar dial tone comes on the line.

Table 9-1. Tone Frequencies Used in Touch-Tone System

Digit	Tone Frequencies (hertz)
1	697 and 1209
2	697 and 1336
3	697 and 1477
4	770 and 1209
5	770 and 1336
6	770 and 1477
7	852 and 1209
8	852 and 1336
9	852 and 1477
0	941 and 1336

3. The number is then dialed.
4. Shortly after dialing, one of two things happens. Either the 20-hertz ringing signal appears on the line, or some sort of busy signal comes on.
5. After the phone being called is answered, the ringing signal stops.

From the above sequence of events, we can see the functions that the automatic dialer must perform. These are listed below:

1. When an intrusion alarm is tripped, the dialer must connect itself to the telephone line.
2. There must then be a delay until the dial tone appears on the line.
3. After the dial tone comes on the line, the circuit must be interrupted the proper number of times in sequence to dial the desired number. An alternative approach is to apply two-tone signals to the line as is done in the touch-tone dialing system.
4. There must be another delay until the phone is answered or a busy signal appears on the line.
5. Finally, the tape-recorded message must be transmitted and the device must be turned off. In some systems, another number is called and another message is transmitted after the first one. In that way, the first call can be to the police department, and the following calls can be to authorized employees who will go to the scene of the intrusion to investigate.

Circuit Operation

Fig. 9-14 shows a block diagram of an automatic telephone-dialing system. The sequence of operation is as follows. When the circuit from the intrusion detector is interrupted, relay K1 is de-energized, closing the contacts that start the system in operation. This connects the system to the telephone line but does not start the tape recorder. An amplifier is connected across the line, and when the dial tone comes on the line, the tape transport is started. The tape has a prerecorded number of tones corresponding to the number to be dialed. These tones are amplified and rectified so that each tone will cause the impulse relay to open, thus dialing the number. A prerecorded tone then causes the tape transport to stop, and it connects an amplifier to the line. This amplifier has a sharp response at 20 hertz so that it will pick up the ringing signal. As long as the ringing signal occurs every few seconds, nothing further will happen. When the ringing signal stops, the tape transport will start and will transmit the desired message two or three times, after which a prerecorded tone will cause the system to shut down.

If the called telephone is not answered, after about one minute a timer will cause the tape transport to start until the next recorded message is reached, which it will then transmit in the same way. In this way, two or three messages can be transmitted. Often, the last message is coded to the telephone operator so that if for some reason the police cannot be reached, the operator can continue to call until she gets a response.

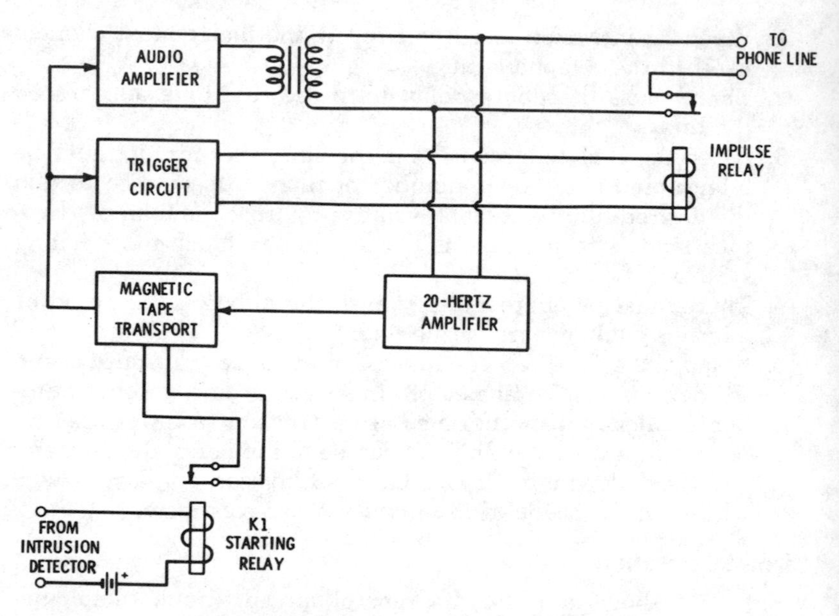

Fig. 9-14. Block diagram of an automatic dialer.

Another feature of the system is an amplifier and filter that will recognize a busy signal. If a busy signal is received after dialing, the tape transport will advance to the next message.

Systems designed for use with touch-tone dialing are somewhat simpler because the impulse relay and its associated amplifier and circuitry are not required. The tones may be recorded directly on the magnetic tape. This is usually done with the telephone set that is also connected directly to the line.

Installation

Connection of an automatic dialer to a telephone is usually accomplished through a coupler furnished by the telephone company. This ensures that the dialer will match the 900-ohm impedance of most telephone lines and that the amplitude will not be great enough to cause cross-talk interference with other telephone lines.

Automatic dialers are often much more complicated than the one shown in Fig. 9-14. They contain additional recorded messages that can be used for other purposes than reporting a burglary. Dial codes are generated by complex electronic circuitry.

Two-Tone Oscillators

One of the system elements of some automatic dialers is an oscillator that will oscillate at two frequencies simultaneously. This circuit is used to generate the two-tone signals that are used in the touch-tone dialing system. Such oscillators can also be used to generate tone codes for other types of signaling systems.

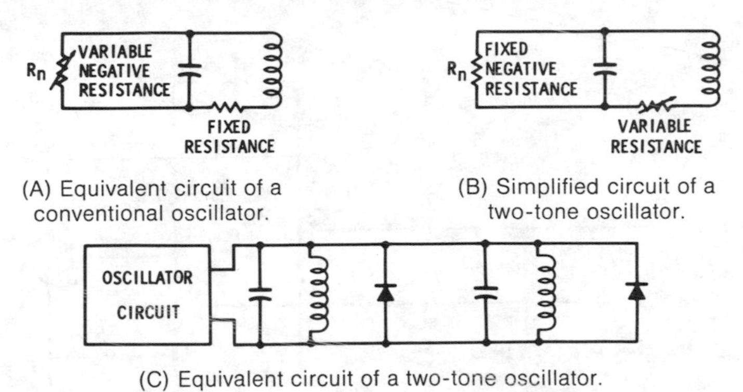

(A) Equivalent circuit of a conventional oscillator.

(B) Simplified circuit of a two-tone oscillator.

(C) Equivalent circuit of a two-tone oscillator.

Fig. 9-15. Oscillator circuits.

The operation of a conventional LC oscillator can be explained using the simplified diagram shown in Fig. 9-15A. The tube or transistor is used to produce an effective negative resistance, R_n. When this resistance is less than the actual resistance R of the tuned circuit, the circuit will not oscillate. If the effective negative resistance is larger than the resistance of the tuned circuit, the circuit will oscillate, but the oscillations will continually increase in amplitude. In a practical circuit, the amplitude increases until it is limited by the nonlinearity of the tube or transistor. This nonlinearity is what keeps the amplitude of the signal from a conventional oscillator constant. However, because of the nonlinearity, the circuit tends to oscillate at a single frequency or to produce spurious signals at several frequencies.

A simplified equivalent circuit of the two-tone oscillator is shown in Fig. 9-15B. Here, the tube or transistor is again represented by a negative resistance, R_n, but this time the value of the negative resistance is kept constant. In other words, the tube or transistor operates in a linear fashion. As we saw before, this would ordinarily lead to an oscillation that would continually increase in amplitude. In this circuit, however, a diode that acts as a nonlinear resistor is connected across the coil of the tuned circuit. As the oscillations increase in amplitude, the current through the diode will increase and its resistance will decrease. This lowers the Q of the coil. Lowering the Q of a tuned circuit is exactly the same as increasing its series resistance. Thus, the oscillations increase in amplitude until the resistance of the tuned circuit is equal to the negative resistance. This limits the amplitude of the oscillations.

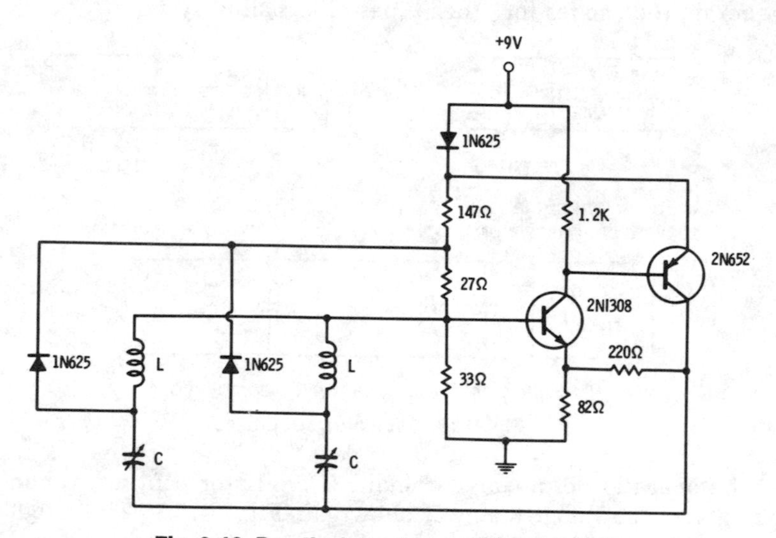

Fig. 9-16. Practical two-tone oscillator circuit.

Since the limiting is done by the tuned circuit, the tube or transistor will operate in a linear manner. We can now use more than one tuned circuit as shown in Fig. 9-15C. Because the circuit is linear, it will now oscillate at two frequencies, just as a linear amplifier can amplify two signals simultaneously.

A practical version of the two-tone oscillator is shown in Fig. 9-16. The two transistors provide a negative resistance in series with the tuned circuits. The diode connected across each coil will limit the signal, and the circuit will oscillate simultaneously at the resonant frequency of both of the tuned circuits. In this circuit, the L/C ratio should be approximately 3,250,000. The values of L and C can then be found from the equations:

$$L = \sqrt{\frac{3.25 \times 10^6}{2\pi f}}$$

$$C = \frac{1}{\sqrt{(2\pi f)\, 3.25 \times 10^6}}$$

where,

L is the inductance in henrys,
C is the capacitance in farads,
f is the frequency in hertz.

USES OF AUTOMATIC DIALING

Automatic dialing devices give small-business operators some of the advantages that they would gain by having their own security guards. In addition, they may be used for many purposes other than protection against intrusion. One of the most important is their use by heart patients to call a doctor if they have an attack. Other less-spectacular, but important, uses permit a businessman to monitor important aspects of his business from his home. For example, a grocer can connect an automatic dialer to a refrigeration system so that it will call him at home if the refrigeration fails. He can then arrange to have repairs made before perishable commodities spoil.

LIMITATIONS

The automatic telephone dialer has some limitations. For example, it may get a busy signal when dialing the police. Most systems are arranged to call again if they get a busy signal the first time, or to place a backup call to someone who will call the police. Nevertheless, there is still a chance that the call will not get through. Unlike systems that use leased lines, it is not practical to equip regular telephone lines with failure monitoring systems that will cause an alarm if a line is cut. It is thus possible for an intruder to disable the system by cutting the phone lines outside the building. The danger from this can be minimized by providing a backup local alarm. Some firms are planning to market radio links to the police department that would avoid this possibility, but these are not yet in widespread use.

False Alarms

Perhaps the most serious limitation of the automatic dialer is the seriousness of false alarms. These systems are not necessarily any more susceptible to false alarms than other systems, but false alarms are much more serious in systems that reach the police department than in systems that merely sound a local alarm or summon a security guard who is already on the premises. The latter calls are annoying and lead to distrust of the system, but they cause no danger to human life. On the other hand, when members of a police department receive a call indicating that a crime is in progress, they usually rush to the scene, risking their lives and the lives of others. They certainly do not take a kindly view toward false alarms. Some police departments report that this has already become a serious problem. Although the manufacturers of automatic telephone dialers do their best to see that they are properly installed, they really have no control of their equipment after it has been purchased by the user. It can be installed by unskilled people, with the result that false alarms may be frequent. It is advisable, and in some cities it is required, that all automatic systems be approved by the local police, who must risk their lives by answering them.

RADIO LINKS

Although radio has been used in security work for many years, its use was largely restricted to police departments. Now, radio is being used increasingly for industrial security. Private security services use two-way radio to dispatch their guards. Many taxicab and bus companies are using two-way radios as a deterrent to robbery and to report robberies quickly. Industries that use radio telemetry to monitor processes such as flow, pressure, and temperature at remote locations are adding additional channels to handle signals from intrusion detectors.

A much more widespread use of radio in security is the use of very small transmitters to extend the alarm system right to the individual. Guards, bank tellers, and cashiers can carry a transmitter no larger than a pack of cigarettes with them at all times. This permits them to trip an alarm without detection.

A transmitter used with alarm systems is shown in Fig. 9-17. This circuit consists of a tunnel-diode crystal oscillator operating in the vicinity of 27 megahertz, with an audio-frequency tone generator. The transmitter has a range of 50 to 250 feet, depending on local conditions. The receiver is simply a communications-type receiver tuned to the frequency of the transmitter. An audio filter in the output is tuned to the same af tone generator.

This arrangement works well with an automatic telephone-dialing system connected as shown in Fig. 9-18. Here the guard, watchman, or bank teller carries the miniature transmitter on his person. When an emergency exists, he merely turns on the switch on the transmitter. The tone-modulated rf signal is picked up by

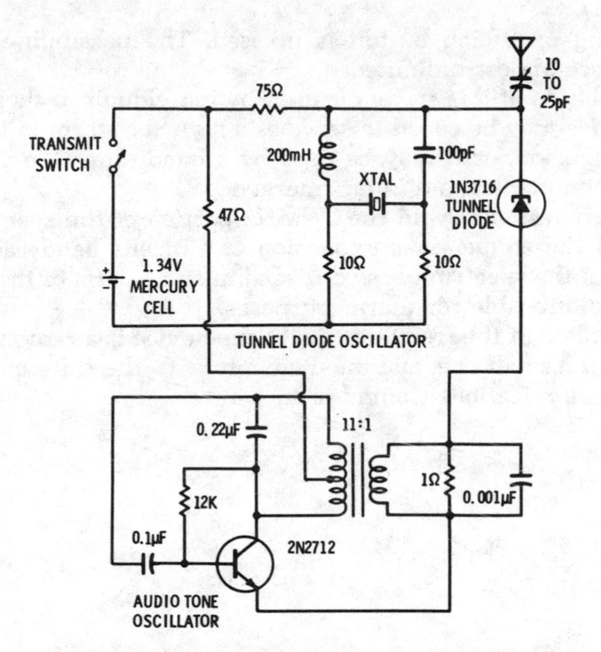

Fig. 9-17. Miniature, 27-megahertz transmitter circuit.

the receiver, which produces an output consisting of the audio tone. This tone is used to turn on the automatic dialer which will then automatically dial and transmit a prerecorded message. If the two-tone oscillator described earlier is used as a tone generator in the transmitter, it is possible to transmit two or more different messages. Thus, it is possible for a guard to carry a miniature transmitter with several different switches. When he discovers an emergency, he can throw a switch and cause a prerecorded message to call for police, firemen, or an ambulance,

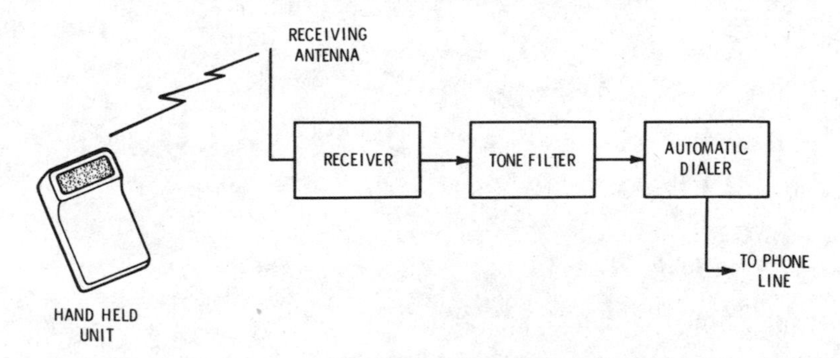

Fig. 9-18. Radio link with automatic telephone dialer.

depending on which button is pressed. The possibilities of this system are almost unlimited.

A variation of this system is used by individuals in their homes. The device may be coded to serve as a security alarm in the event of an intrusion, or it may be used by a handicapped or ill person to summon medical aid in an emergency.

Modern systems avoid the 27-MHz portion of the spectrum because of the tremendous expansion of Citizens band radio. This portion of the spectrum is so crowded in some parts of the country that it is unusable for alarm purposes.

Regardless of the frequency used, the device is a radio transmitter, albeit a small one, and must comply with the rules of the Federal Communications Commission.

CHAPTER 10

Accessories and Auxiliary Equipment

There are many components in electronic security systems that can be used with any of the intrusion detectors described in earlier chapters. These include power supplies, access switches, monitoring equipment, event recorders, and cameras. These accessories can be used to increase the versatility and effectiveness of any electronic security system.

ACCESS SWITCHES

Any intrusion alarm system must have a provision for allowing authorized persons to enter the protected area without setting off the alarm. This is usually accomplished by some sort of switch that will disable the alarm system.

The access switch is actually one of the most important links in any security system. What good is even the most sophisticated system if an intruder can locate and operate the access switch to turn the system off? Yet, many burglaries have been carried out in just this way.

Key-Operated Switch

The most common type of access switch is a key-operated switch (Fig. 10-1) similar to the ignition switch of an automobile. Keys for the switch are issued only to authorized personnel. Additional protection is sometimes provided by using two different access switches that require different keys to operate them. In this way, the keys can be issued to two different employees, so that both employees must be present to shut off the alarm. No one employee may enter the facility alone without setting off the alarm.

The access switch may be connected to remove power from the entire alarm system, or it may disable the alarm only. The exact arrangement depends on the type of detector and alarm system.

A great deal of sophistication is used in the manufacture of keys for access switches to prevent their being counterfeited or the lock being picked. Sometimes a vibration detector is included with the

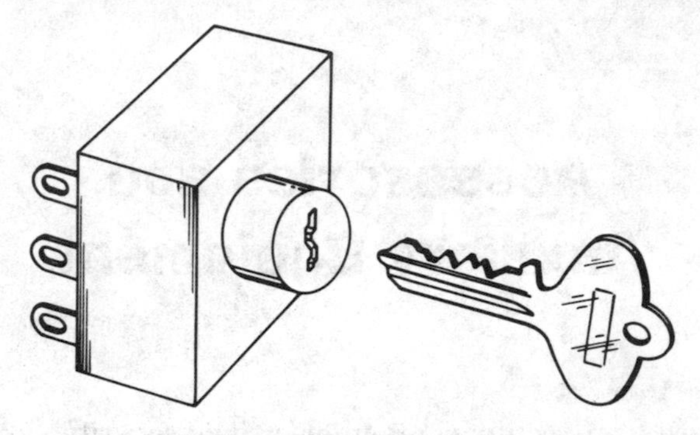

Fig. 10-1. Key-operated switch used for access control.

lock. Its sensitivity is set so that it will not be tripped by normal operation of the lock, but it will trip an alarm if the lock is treated roughly as it might be if someone were trying to pick it. Additional protection is often provided by using a more complex arrangement than a simple single-pole, single-throw switch. In the circuit shown in Fig. 10-2, one set of contacts must be closed and the other opened to disable the alarm system. If an intruder tries to foil the alarm by jumping or opening wires, he will not succeed unless he gets the right combination.

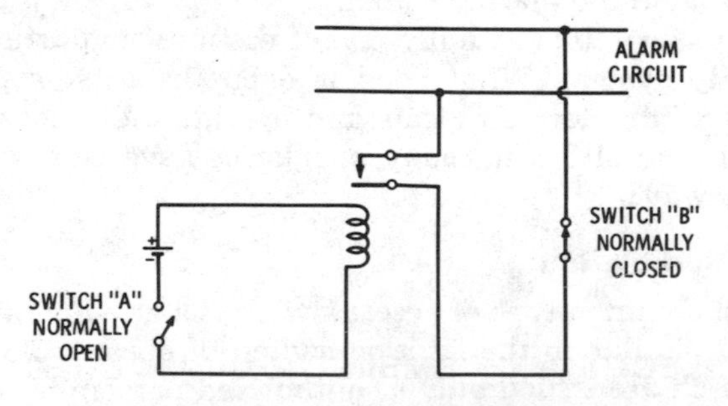

Fig. 10-2. To short the alarm circuit, switch A must be open and switch B must be closed.

Disadvantages

The chief disadvantage of key-operated access switches is that the keys may be lost or stolen. Anytime a key is missing even for a short time, the lock should be changed, because a clever intruder can have a duplicate key made very quickly.

TIME SWITCHES

One of the limitations of an access switch that can be operated only by a particular person is that a serious intruder who thinks that he has a lot to gain by the burglary can kidnap the person or a member of his family and force him to open the premises and disable the alarm. Many burglaries have been performed this way, and where the employee has resisted, he has been seriously injured.

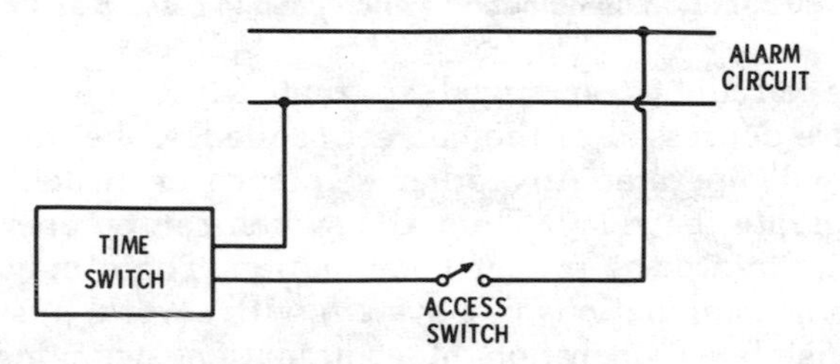

Fig. 10-3. An access switch that will disable alarm only at certain times.

A system that will eliminate this possibility is shown in Fig. 10-3. Here, a time switch is connected in parallel with the access switch. For the alarm system to be made inoperative, both the key-operated access switch and time switch must be opened. The time switch is set to open during normal business hours. Thus, if someone is forced to open the key switch, the alarm system will still be active. Time switches are available with "omitting devices" that can be set to prevent them from operating on certain days of the week, such as Saturday and Sunday, or a day when the protected establishment is not normally open. This is an important feature; otherwise, a burglar might force a store manager to open the store during normal business hours but on a day when it is closed, such as on a Sunday.

ELECTRICAL COMBINATION SWITCHES

Electrical combination switches such as those shown in Fig. 10-4 are becoming increasingly popular as access switches for secure

Fig. 10-4. Push-button combination switch used to disable alarm circuits.

areas. The circuit is arranged so that when the push-button switches are depressed in the correct sequence, the alarm-disabling relay will operate. Any other sequence of switch operation will not operate the relay. Thus, the switch can be operated only by one who knows the proper combination. There may be up to 30,000 wrong combinations in a system with several push buttons.

The principle of operation of combination switching systems may be understood by first considering the latching relay shown in Fig. 10-5. In this circuit the relay is energized by momentarily closing switch S, which may be a simple push button. Once the relay is energized, contacts A will close, holding the relay in the energized state after switch S is opened. The other contacts, 1, 2, and 3, may be used for logic functions.

In the circuit of Fig. 10-6, the latching relays are arranged so that the contacts of relay No. 2 supply power to the coil of relay

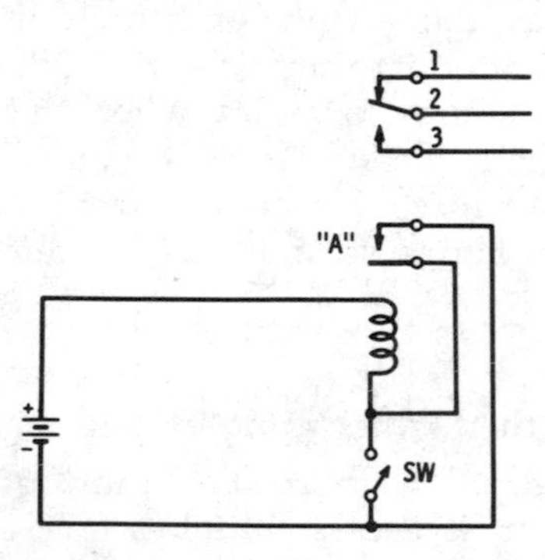

Fig. 10-5. Latching relay circuit.

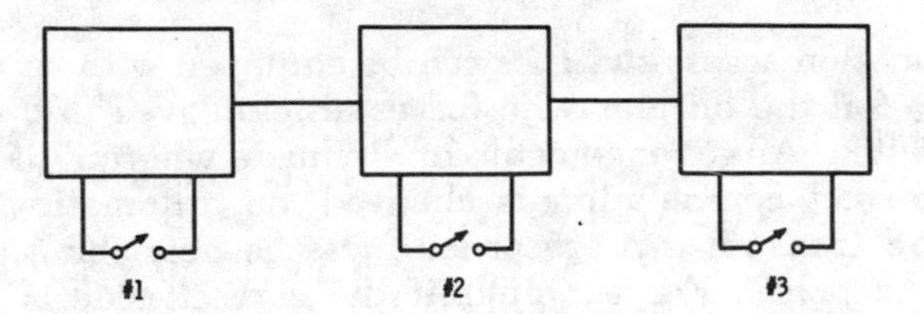

Fig. 10-6. Basic principle of combination access switch.

No. 1, and the contacts of this relay in turn furnish power to the circuit of relay No. 3. Thus, the push-button switches must be pressed in the order 2, 1, 3. Pressing the buttons in any other sequence will have no effect. An additional refinement that may be added will de-energize all of the relays if a button is pressed out of sequence. When the buttons are pressed in the proper order, the last relay in the chain will disable the alarm system.

This arrangement has the disadvantage that an intruder may keep experimenting until he happens onto the proper sequence, although if several different circuits are used, the odds are against this. A simple addition that will overcome this limitation is shown in Fig. 10-7. Here, a line from each of the buttons is connected to an OR circuit. This circuit will provide an output whenever any one of the buttons is pressed. This output drives a trigger circuit that applies a small pulse to capacitor C, which discharges slowly through resistor R. The discharge rate of the RC circuit is set so

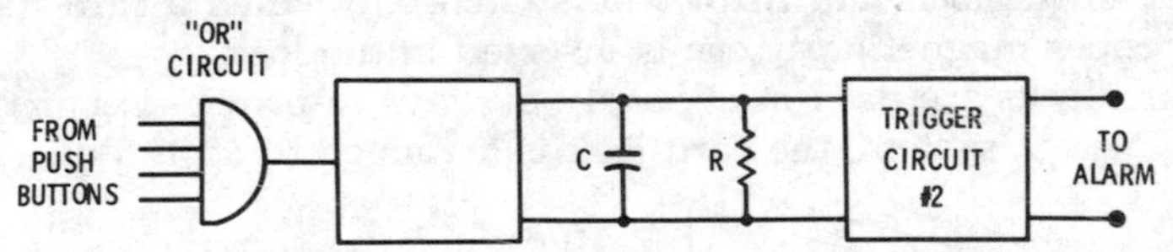

Fig. 10-7. Circuit to set off alarm when push-button access switch is tampered with.

that if all of the push-button switches are pressed only a few times in a ten-minute period, no substantial voltage will develop across the capacitor, and the second trigger circuit will not operate. Thus, if someone operates the switches in the proper sequence, the second trigger circuit will not do anything. Even if he makes a mistake and has to try again, the voltage across the capacitor will still be small. If, however, an intruder should start pressing the switches at random in an attempt to break the code, the voltage across the capacitor will soon build up enough to operate the second trigger circuit, which will then either initiate the intrusion alarm or activate a signal at a central location, thus indicating that someone is tampering with the combination switch.

Combination access switches can be equipped with an ingenious device to foil the burglar who forces an employee to operate the access switch. An arrangement can be made whereby if one digit in the correct combination is changed, an automatic telephone dialer will transmit a prerecorded message describing the situation to the police. For example, if the correct code is 7819, the number 7818 may be arranged to dial the police. Thus, if a burglar, knowing that a particular employee has the combination, obtains it under duress or forces an employee to use it, a message will go to the police that someone is being forced to open the facility. The message may include the warning that the employee or his family may be in danger if the police appear openly. In this way, the police can act accordingly and apprehend the burglar at the opportune time.

CARD-LOCK SYSTEMS

Another type of access control system used frequently in connection with intrusion alarms is the card lock shown in Fig. 10-8. In this arrangement, a coded card is inserted into the lock to throw the switch. Any card that is not properly coded will have no effect.

Cards can be coded in many different ways. One effective way to code cards is to deposit small amounts of magnetic material in the card in such a way that they cannot be seen. A magnetic detector in the lock will throw the switch only when a card having the proper magnetic pattern is inserted in the lock.

Card locks are used in a variety of ways to provide security. In the simplest system, the card lock is arranged to open a door lock

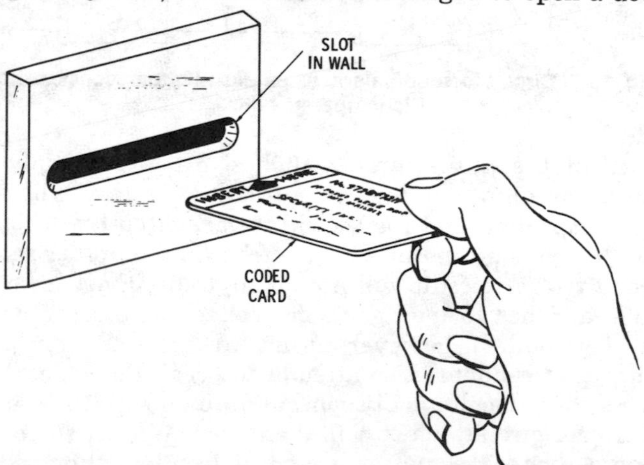

Fig. 10-8. Coded-card switch used for access control.

whenever the proper card is inserted in the lock. This provides a moderate degree of security in allowing only authorized persons to enter an area. Uses vary from a private club where it is desired to admit only members, to an industrial plant where additional protection for a security area is needed.

The principal advantage of the card lock is that it is simple and inexpensive. When properly used in connection with other measures, it will contribute substantially to the security of an installation. Its principal limitation is that cards may be lost or stolen and used illegally before the combination can be changed and new cards can be issued.

HOLDUP ALARMS

Any intrusion alarm can also be used as a holdup alarm, and many systems have this provision. This is accomplished by disconnecting the intrusion detection system and connecting a switch or series of switches that can be used to trigger the alarm in the event of a holdup during normal business hours. There are many different ways in which switches can be hidden so that they can be actuated without attracting the attention of the holdup man.

One of the most important requirements of a holdup alarm is that it be capable of being actuated without attracting the robber's attention. Most security people feel that the danger of irritating a criminal to the point of harming a person is very great if he knows who sets off the holdup alarm. In banks or ticket offices where the

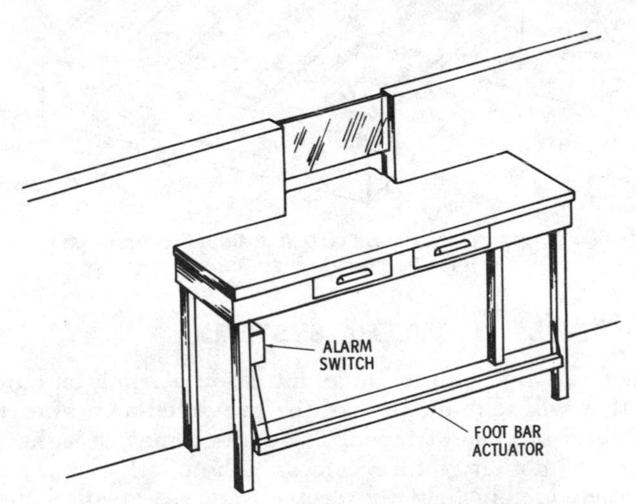

Fig. 10-9. A foot-operated switch which permits the intrusion alarm to be used as a holdup alarm.

lower part of the cashier's body is hidden from view, an alarm can be tripped secretly by the use of a foot-bar switch such as that shown in Fig. 10-9. In stores where the employees can be watched by the robber, this is not advisable.

The money clip shown in Fig. 10-10 is an ingenious holdup switch that may also be used in a regular intrusion alarm as a detector. This switch will be actuated whenever anyone removes the money from the clip. The switch can be arranged so that it is cut out of the circuit by a hidden push button. Thus, whenever an authorized person takes money from the clip, he will first disable the alarm by means of the hidden push button. A thief, however, not realizing the way the system operates, may simply remove the money and unwittingly trip an alarm that will inform the police what he is doing.

Many other types of holdup alarms will suggest themselves for particular applications. The principal consideration is that the switch must be located so that it can be actuated without attracting any atention, or so that the holdup man will unwittingly actuate it himself.

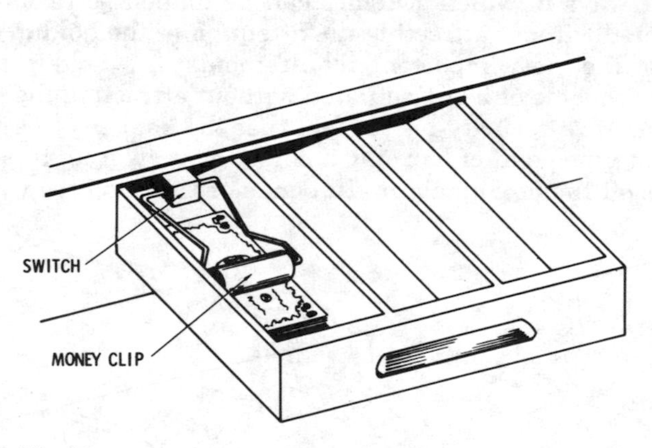

Fig. 10-10. A money clip that will initiate the alarm when it is lifted higher than normal.

CAMERA SYSTEMS

Another valuable addition to an intrusion alarm is an automatic camera that will take pictures of the protected area after the intrusion alarm has been tripped. The camera may be either a still camera or a motion picture camera. Photoflash lamps may be added to adequately light the protected area. Perhaps the most effective arrangement is a camera that is timed to take a picture every few seconds after the intrusion alarm has been tripped.

The camera system will add to the security of an installation in three ways. First, the fact that his actions are being photographed will often frighten an intruder into fleeing without stealing anything. Secondly, the photographic record will aid in identifying the criminal, and lastly, the photographic record of a completed crime will certainly point the way to changes that will make the area more secure.

EVENT RECORDERS

Many thefts from large plants and institutions are carried out with the assistance of guards and other trusted employees. Many security people feel that a means must be provided to check on trusted employees if for no other reason than to remove the temptation to steal or to collaborate with thieves. One important accessory that can help in this function is an event recorder of the type shown in Fig. 10-11. In a large security system, the event recorder is connected to record as many different security functions as is deemed necessary. For example, it may be arranged to print out a record giving the date and time of every change in a security system. It will make a record every time that an access switch is thrown from secure to access, every time an intrusion alarm is tripped, or every time a door or gate is opened outside of normal business hours. A record of this type, shown in Fig. 10-12, has many functions. It provides an excellent opportunity for checking the cause of false alarms. It was mentioned earlier that one scheme for frustrating the most elaborate intrusion alarms is for a prospective burglar to cause a number of false alarms without being

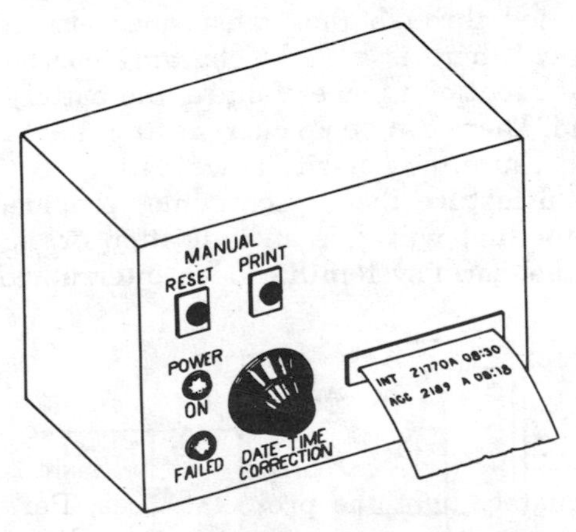

Fig. 10-11. An event recorder which provides a record of all intrusions and false alarms.

detected. After the number of annoying false alarms has been so great that no one has any faith in the alarm system and will no longer respond to it, the intruder quickly loots the protected area. If a record is kept of all false alarms and their cause is carefully investigated, this practice can be spotted quickly. In this case, the area in question will actually be given additional surveillance, and the intruder will be caught.

Another valuable function of the event recorder is that it will provide a check on security guards. For example, if a security guard at a central monitoring point intentionally allows a thief to enter a factory gate after hours, a record will be made and he can be tracked down.

Many other supervisory uses are often made of event monitors. They can keep a record of the times that watchmen make their rounds, of fire alarms or other emergencies, and of any other important plant functions that it is desirable to monitor.

```
GATE B     ACC SEPT 6    1:18PM
SECTION A  ALM SEPT 6   11:00AM
```

Fig. 10-12. Typical event recording of changes in status of a security room.

POWER SUPPLIES

An intrusion alarm is only as good as its power supply. If an intruder can disable an entire alarm system by merely cutting the power lines, the alarm has a very limited value. The most popular type of alarm for intrusion alarms is an ac supply with a battery floating across the dc output, as shown in Fig. 10-13. Normally, power is supplied through the transformer and rectifier to the load. In addition, there is a slight charging current through the battery. In the event of a power failure, the battery supplies current to the load. There can be no current from the battery through the rectifiers because the polarity is wrong.

An additional feature that is sometimes provided is a power-failure indicator that will give an indication or alarm whenever the power to the alarm system has been interrupted. This can be

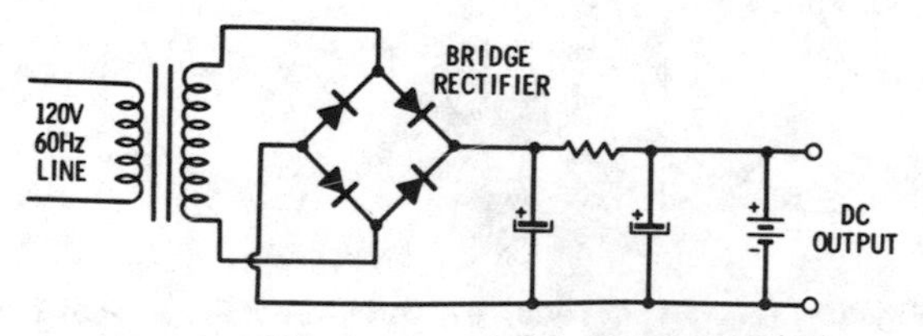

Fig. 10-13. Typical intrusion-alarm power supply.

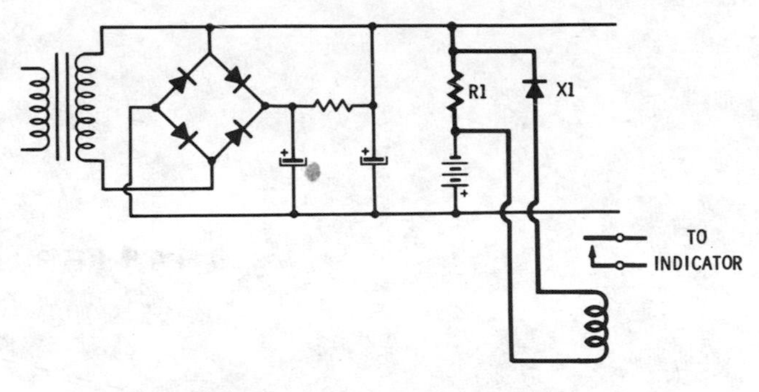

Fig. 10-14. Power supply with a power failure indicator.

accomplished by the circuit shown in Fig. 10-14. This circuit is the same as that of Fig. 10-13 except for the power-failure alarm shown in heavy lines. Normally, when the ac power is operating properly, there is a small charging current through the battery and resistor R1, which has a value of a few ohms. The direction of current is such that the top of the resistor is positive. There can be no current through diode X1 because it is connected so as to oppose the normal direction of current. When the ac power fails for any reason, the situation changes. The battery now furnishes current to the load, and the direction of current through resistor R1 is reversed. The bottom end of resistor R1 is now positive, and current goes through diode X1 and the power-failure relay. The power-failure relay will now be energized and will cause an alarm or indication whenever the ac power to the intrusion alarm is interrupted.

With this arrangement, a potential intruder cannot cut the power lines to the intrusion alarm one day and then wait a few days until the battery has discharged before attempting a burglary.

Power-failure alarms can also be used for purposes other than merely protecting an intrusion alarm. For example, additional contacts on the relay can be used to turn on emergency lighting systems.

CHAPTER 11

Practical Applications

The wide variety of electronic intrusion detectors that are available makes it possible to provide a high degree of security for almost any type of building. At the same time, this variety complicates the problem of determining just which system is best suited for a particular application. This is not an easy problem to solve. It requires not only a thorough knowledge of the advantages and limitations of the various systems, but a good knowledge of security principles in general and the methods that a burglar might use to enter a particular building. Many firms have found that it is advantageous to use the combined talents of electronics and security people to provide the best system for a particular building area. In this chapter, the general principles of application of electronic security systems are outlined and examples of specific installations are given.

The first factor to be considered in any application of security electronics is the economics of the situation. Until the property owner or manager feels that the cost of a system is justified, little can be done. In assessing the cost of providing electronic protection, two major factors must be considered—the risk involved and the effect on insurance rates.

EVALUATION OF RISK

What a business stands to lose by a burglary is often much more than the cash value of any goods that might be stolen. Frequently, a great deal of property is damaged when burglars break into a place. Where doors and windows are protected, the burglars might cut through the floors, walls, or ceiling. Although the value of the stolen goods and the damaged property may be covered by insurance, other losses may not be. A firm might lose a great deal of business as a result of not having the stolen items in stock after the burglary. Another possible loss that is often not considered is the loss that would result from careless or malicious destruction of the records of a firm. Such a loss in the long run might be much more costly than the loss of merchandise.

INSURANCE RATES

Most businesses carry at least some insurance protection against burglary and theft. When an electronic security system is installed or an existing system is improved, it is a good idea to check with the insurance company to see if a more favorable insurance rate can be obtained. Underwriters Laboratories have issued standards covering various types of electronic security systems. Usually, a system must meet these standards if it is to qualify a property for more favorable insurance rates.

The insurance company can often provide valuable guidance in selecting security systems for a particular type of business. These companies usually employ experts who are familiar with all types of systems, and in addition they have had a great deal of practical experience with burglaries.

GENERAL SECURITY MEASURES

When planning the details of an electronic security system, one must not forget that an electronic system will not eliminate the need for other common-sense security measures. An electronic system will be most effective in preventing intrusions, or in leading to the apprehension of an intruder, if other factors combine to maximize the time required to enter the premises and steal anything. Even the most advanced security system will be ineffective if a burglar can break into a place, trip the alarm as he does, and escape before anyone has time to respond to the alarm.

The first line of defense against intrusion is good fences, walls, gates, and doors with locks that are not easily picked or broken. All of these will increase the time required to commit a burglary. Along these same lines, cash and small items of merchandise that may be readily converted to cash must be secured as well as possible. Cash should be kept in a good safe that cannot be easily cracked or carried away. Larger items that can be quickly converted to cash, such as automobile tires, should be secured, in this case by a chain and padlock.

Another deterrent to burglary is good lighting. A burglar is less apt to spend a lot of time opening a well-lighted safe that is clearly visible from a front window, than one that is hidden from view.

One common-sense security measure is often completely ignored: nothing should be left lying around that a burglar could use to break into a building. There are many cases on record where burglars have used heavy boxes that they found on the premises to break a window, or have used ladders that they found behind a building to climb to the second story, where they could gain access to the interior.

Finally, it should be realized that there is no electronic substitute for a security guard on the premises. For complete effectiveness, all larger industries should have security guards as well as electronic systems. There is no substitute for the human judgment of a guard in recognizing false alarms and deciding on the most effective course of action in the event of an actual intrusion.

SELECTING AN ALARM AND SIGNALING SYSTEM

The particular type or types of alarm and signaling systems that should be used in a given application depend on many different factors, such as whether the buildings are located in an area patrolled by police, in a remote area, or somewhere in between. In general, the purpose of an alarm is to obtain a response that will stop the crime or lead to the apprehension of the burglar. It is important that the response come in the shortest possible period of time. There are four types of alarm and signaling systems that can be used:

1. A private security company
2. A centralized alarm
3. A wired or automatic dialer system
4. A local alarm

Using the Private Security Company

When the head of a firm decides to use the services of a private security firm, he turns the whole problem of security over to them. They select the intrusion detectors that they consider necessary to provide the particular type of protection called for in the contract. This can range from perimeter protection to protection of a specific object such as a safe. The degree of protection required should be carefully explained in the contract. The cost of the service depends on the degree of protection required.

The use of a private security company has certain advantages. In addition to alarm systems, the company maintains patrols who are apt to spot any burglar who manages to foil an alarm system in the protected area. The intrusion detectors installed by the company are connected by telephone wires to their headquarters. They can then exercise some judgment before calling the police. The cost of private security services varies widely with different areas. The types of service rendered by private security companies are rated by Underwriters Laboratories.

The Centralized Alarm System

In any establishment where a security force is available, it is customary to connect all intrusion detectors to a central location, usually the security guards' headquarters. This system makes it possible for a smaller number of guards to provide a greater degree of security. A further advantage is that the use of time clocks and event recorders provides a check on the guards themselves.

The Automatic Telephone Dialer

The automatic telephone dialer is becoming very popular because it provides the small firm with a system that will summon the police the minute an intrusion occurs. On the surface, it appears as an ideal solution to most security problems. The wide use of such systems, however, detracts from their effectiveness. In some large cities, the police are receiving so many false alarms from automatic dialers that they are assigning a low priority to such calls. As mentioned previously, in some cities, ordinances prohibit direct dialing of the police department by automatic dialers. Perhaps the primary reason for this is the large number of false alarms. For example, in one city, over 90 percent of the calls received from automatic dialers proved to be false alarms.

At the very best, a police car rushing to the scene of a crime in progress jeopardizes the lives of the officers and other motorists. When such calls are false alarms, this represents a needless and careless risk.

In spite of these limitations, the automatic telephone dialer definitely has its place, and is probably the best type of signaling system for many industries. Perhaps the best type of automatic dialing system consists of two intrusion detectors that are arranged to transmit messages of different priorities. The first could be connected to some type of perimeter protection system so that when the alarm is tripped, it will dial the police and report that there is an indication that someone is trying to enter the premises. The police could then dispatch a car to the scene at safe and reasonable speeds. The second message would be sent when the burglar had progressed to some other area, and would be of a higher priority. The police would be reasonably sure that a real intrusion was taking place if, for example, they were to receive three successive messages indicating that a burglar had first entered the grounds, had then entered an inner area, and finally was tampering with a safe. On the other hand, if only one of these alarms were tripped, they would recognize the possibility of a false alarm.

In any event, an automatic dialer should not be installed without a conference with the police department that will be expected to respond to it.

The Local Alarm

In most applications, a local alarm is not considered very effective unless it happens to be an area that is patrolled regularly by police. In this day of noninvolvement, it is not likely that a public-spirited neighbor will respond to an alarm.

There are, however, cases where the local alarm is the only practical type to use. These include homes and businesses that are so remotely located that neither the police nor a private security service could reach the scene in a reasonable period of time. The local alarm has a definite psychological effect on all but the most intrepid intruder. When combined with other devices such as flashing lights, it can be made very effective in scaring burglars away before they have completed their burglary.

Another good place for a local alarm is in a private home where the resident feels fully capable of defending himself against intruders, but needs a system to wake and warn him that an intruder is in the area.

SELECTING AN INTRUSION DETECTOR

The problem of selecting intrusion detectors is complicated simply because there are so many different types from which to choose. Checking over the advantages and limitations of the various types described in the preceding chapters will show that there are some obvious applications where one or more types would not be acceptable, but usually this will still leave a choice of two or three different detectors for each application.

Perhaps the best way to determine the detector that is best suited to a particular application is to prepare a survey similar to that shown in Table 11-1. This survey will first determine whether perimeter, area, or spot protection is required. Next, the conditions that prevail in the area to be protected will be determined. These can be checked against the advantages and limitations of each of the detectors described in earlier chapters. In this way, a system, or combination of systems, can be selected that will provide optimum protection.

Type 11-1. Survey for Intrusion-Detector Installation

I. Type of Protection Required	
Perimeter	________
Area	________
Object	________
II. Number of Entrances to Area	
Door	________
Windows	________
Emergency exits	________
III. Environmental Conditions	
Temperature range	____ to ____ degrees
Humidity range	____ to ____ percent
Ambient sound sources, outside	
traffic	________
trains	________
bells or whistles	________
Ambient sound sources, inside	
radiator, heater, etc.	________
fans, air conditioner	________
machinery	________
Moving objects in protected area	
fans, machinery	________
loose doors, windows, etc.	________
Radio-frequency conditions	
nearby radio or tv stations	________
machinery that will generate rfi	________
equipment susceptible to rfi	________

TESTING AN INSTALLATION

After an installation is completed, it must be thoroughly tested. The best test is to act like an intruder and try to foil the system. Cut the power to the system and cut any exposed wires. If the system can be foiled, it is only a matter of time until an experienced burglar will foil it. Many systems that respond to rapid motion will not be affected by very slow motion. Try to walk through the protected area very slowly. Do not neglect the fact that the "stay-behind" can usually find a place to hide and will not trip the perimeter system until he leaves the premises.

ESTABLISHING PROPER PROCEDURES

The effectiveness of any alarm system depends on how well it is used. Definite procedures must be established and responsibilities must be assigned. The number of people who know the details of the security system, including how it works and where the wiring is located, should be kept to an absolute minimum. To make maximum use of an electronic system as a deterrent, tell as many people as possible that it is installed and as few as possible how it works. The wiring must not be shown on any building plans that might fall into the hands of a potential intruder. By the same token, an alarm system should not be installed when a building is erected, but later. Most banks set a good example in that respect. Everyone knows that the bank has an alarm system, but very few people know just how it works or where the various parts of the system are located. Another consideration in this respect is control of the number of persons who can operate the access switches. For a system to be effective, access must be limited to a minimum number of persons.

A security system must be operated in such a way that it provides adequate security at all times. In many installations, it is the usual practice to shut off the entire system as soon as the first authorized employee arrives at work in the morning. This means that the back door and emergency exits are no longer protected. In many cases, merchandise has been taken out through an unprotected back door soon after the first employee arrived at work and disabled the security system. If the system had been properly installed, the back doors would have remained protected until later, when more supervisory people were at work. Some firms have found it advisable to keep emergency exits, which are not normally used, protected at all times.

One important procedure that should be established and rigidly enforced is the investigation of every false alarm. There are several reasons for this practice. First, false alarms indicate that the system is not reliable and should be improved. Secondly, and equally important, is the fact that many false alarms are caused by potential intruders trying to learn about the intrusion detection system, or to undermine faith in the system.

If an intrusion protection system is worth installing, it is worth keeping in operation, and in good operating condition. A casual inspection of the protective foil of many business establishments will show that the system is inoperative and probably has not been used for months. This is an open invitation to the skilled burglar to break in.

EXAMPLE OF SMALL-STORE PROTECTION

The following is one example of a protection system successfully used in a small store. A variety store contained many items of value, including a large amount of cash. Its location was so remote that the quickest response that could be counted on was about twenty minutes. For this reason, management decided that a local alarm that would upset a potential burglar was the best protection that could be provided. Most of the items in the store were bulky, and the smaller items could be put in the safe at night, so it was felt that any burglary attempt would be aimed at the safe. The safe was located where it was plainly visible from the front window of the store, and a vibration detector was installed in the store. In order to prevent burglars from kidnapping the manager of the store and forcing him to disable the system, a time switch was installed. This prevented anyone, even authorized employees, from disturbing the safe except during normal business hours. The alarm was connected to a loud klaxon horn aimed at the protected area. The sound level from this horn was so loud that it was actually painful, as well as being audible for several blocks. Combined with the horn were photoflash lamps that flashed blinding light into the protected area. Since installation of the system, the store has not had a successful burglary.

EXAMPLE OF AN OFFICE PROTECTION SYSTEM

The following is another example of a successful protection system. In this case, the protection system was installed in an office located in a public office building. There was some control over the people entering the building after hours, but it certainly was not very rigid. Furthermore, the cleaning people employed by the building had keys to the offices so that they could clean after regular office hours. The valuables to be protected included a small amount of cash, valuable records, office furnishings, and files containing proprietary information that could result in the loss of competitive bids if it were to fall into the hands of competitors.

The problem was turned over to a private security company. They put a switch on the door and protective foil on the windows. These were wired to an alarm at the headquarters of the security agency. Of course, the alarm was tripped every time that the cleaning people entered the area, but an arrangement was made that none of the cleaning people would enter the area after 10 P.M. If any of the employees had reason to enter the area after this hour, they had to call the security company and make arrangements, including giving a secret code word. To provide additional protection, the valuable records and cash were kept in a heavy safe protected by a vibration detector which was wired to the headquarters of the security agency. If this detector were jarred momentarily by the cleaning people, the alarm would be ignored, but if it were disturbed continually, the police would be called. The arrangement has proven highly satisfactory for a long period of time.

CHAPTER 12

Electronic Eavesdropping and Secure Communications

All of the electronic security systems described in the preceding chapters are used to provide security against vandalism, personal injury, or theft of property. There is another aspect of security in which electronics is playing an increasingly important role. This is the area of electronic "eavesdropping" or "bugging," as it is commonly called. Illicit eavesdropping is being carried on by electronic means for a variety of reasons. Legitimate law-enforcement agencies use bugs to capture criminals, and private investigators use them to obtain information for their clients. The most significant application of electronic eavesdropping devices is their use for stealing business or personal secrets, or for listening to police broadcasts by criminals who are trying to avoid apprehension.

There is no way of telling just how much illicit eavesdropping is being carried on. Several companies are engaged in the manufacture of these devices, but much of their output goes to legitimate law-enforcement agencies.

We do know, however, that anyone with even a rudimentary knowledge of electronics can build a miniature bug that can be easily concealed. In fact, with the advent of microminiature circuitry and compact batteries, the state of the art seems to be on the side of the eavesdropper, rather than on the side of the law-abiding citizen. The availability and ease of construction of electronic eavesdropping devices make it safe to assume that any given installation may be bugged if someone considers it worth the risk and effort. For this reason, wherever business secrets are considered worth stealing, it is advisable to take adequate countermeasures against eavesdropping. Several management consultants now specialize in the "debugging" of business facilities, particularly those business areas where information of an important and secret nature is discussed.

Another area in which secure communications are needed is in the computer field. The computer has proven to be such a valuable aid to business that it is being used to handle all sorts of information, much of which is of a confidential and secret nature. In some firms, all of the records are kept in computer memories. If a competitor gains access to the computer, he could obtain information that might destroy the victimized company.

In addition to stealing information from a computer, a competitor can change the information stored in the computer, once he has gained access to it. Many crimes against business consist of juggling the records in the computer in such a way that the criminal benefits financially from the change.

Computers obtain their information through terminals. These terminals are connected to the main frame of the computer either through direct dedicated wires or through a communication system such as the telephone company. In the latter case, unless security precautions are taken, it is possible for anyone to dial the proper number and gain access to the computer. Once access is gained, the criminal may either steal the information or change the data, either for his personal gain or as an act of sabotage to harm the host company.

Techniques to improve the security of data communications are discussed later in this chapter.

BUGGING OR EAVESDROPPING DEVICES

There are two areas of interest in eavesdropping devices. Many law enforcement agencies need information on techniques that they can use in the detection of crimes and the apprehension of criminals. Law-abiding citizens and security personnel need information on the subject so that they can protect themselves against the eavesdropper.

Fortunately for our purposes, the same information is required for bugging and for countermeasures. We must know how the bugs work if we are to use them intelligently, or to find and disable them.

Before discussing the various countermeasures that can be taken against eavesdropping, we will briefly review the principles of operation of some common eavesdropping devices.

Most of the eavesdropping devices that are in use today fall into one of the following categories:

1. Hidden microphones connected by wires to a listening or recording point.
2. Hidden radio transmitters.
3. Telephone wire taps.

HIDDEN MICROPHONES

The hidden microphone is simply what the name implies, a miniature microphone that is hidden some place where it will not be noticed and that is connected by concealed wiring through a sensitive amplifier to headphones or a tape recorder. Microphones may be made very small, particularly where high amplification is available to recover the signals. The use of integrated-circuit amplifiers makes large amounts of amplification available in a very small space. The power required to operate these miniature amplifiers is extremely small. They can operate continuously for weeks or months on power from a small battery.

The "Spike" Microphone

Miniature microphones may be concealed in a variety of places, such as picture frames, ashtrays, lamps, etc. An interesting eavesdropping microphone is the "spike mike" shown in Fig. 12-1. This device consists of a small microphone element, similar to a phonograph pickup mounted on a spike, together with a miniature amplifier. The spike can be driven into a door or wall of the area to be monitored, as shown in Fig. 12-2. The spike carries the sound vibrations to the microphone in much the same way as a phonograph needle carries sound to the pickup element. Although the sound picked up by a spike mike is not as clear as that from a regular microphone, it can produce remarkable results. Most modern spike mikes use integrated-circuit amplifiers.

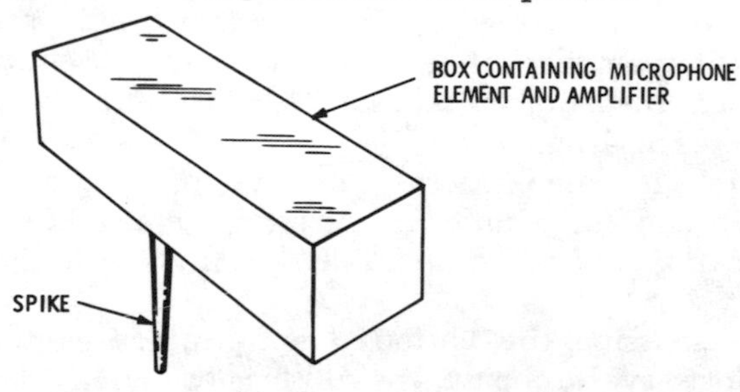

Fig. 12-1. Microphone mounted on a hollow spike.

Fig. 12-2. Spike mike driven in the wall to eavesdrop on a conversation.

The ability to plant and conceal a microphone is limited only by the ingenuity of the eavesdropper. The wiring is usually more difficult to hide. When low-impedance microphone elements are used, very fine wires may be run around the walls or under carpets, or they may actually be imbedded in the carpets.

Directional Microphones

A variation on the hidden microphone is a high-gain directional microphone that may be used to pick up conversations up to 300 feet away. In order to use a microphone to pick up a conversation many feet away from its location, the microphone must have a

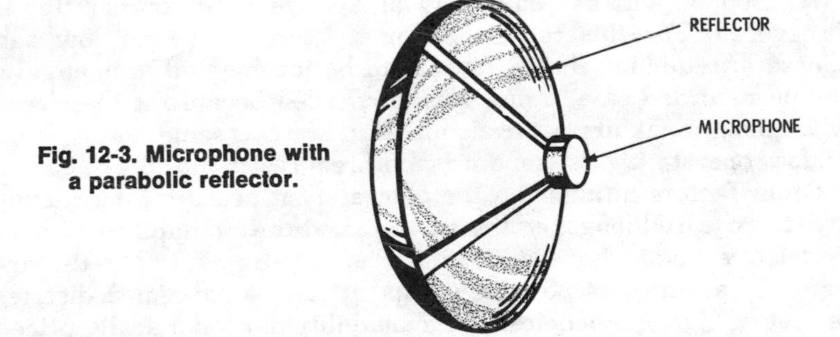

Fig. 12-3. Microphone with a parabolic reflector.

very narrow beam. Otherwise, the extraneous sound reaching the microphone will completely mask the conversation. Fig. 12-3 shows an arrangement that has a sound pickup confined to a narrow beam. Here, an ordinary microphone element is mounted at the focal point of a parabolic reflector. The reflector focuses the sound waves arriving from a single direction on the microphone, just as a parabolic mirror focuses light. With a reflector diameter of 3 feet and a high-gain amplifier, spoken conversations may be monitored at distances ranging from 150 feet to over 300 feet, depending on the amount of background noise.

A disadvantage of the directional microphone is that it is large and may be spotted easily if used during daylight hours. It is most successfully used at night or concealed behind an open door or window.

HIDDEN RADIO TRANSMITTERS

By far the most popular type of bug in use today is the miniature radio transmitter. This is simply a very small transmitter concealed in the area to be monitored. It picks up the conversations in the monitored area and transmits them to a remote monitoring area. The range of these devices is a matter of compromise and is usually limited by the permissible size and available power. If the transmitter can be hidden in a radio, television set, or lamp, where ac power is available and size is not a serious limitation, it can operate continuously and may have enough power to transmit a mile or more, but this is uncommon. The usual range of such a transmitter is about 300 feet.

Operating Frequency

Theoretically, the eavesdropper can use any frequency that he wishes for his clandestine transmitter. For practical reasons, however, only a few frequencies are suitable. In the first place, the eavesdropper will be completely ineffective if his transmitter is discovered. For this reason, he must keep the power low and choose a frequency that is not apt to be intercepted accidentally. In one recorded case, a bug was discovered because it interfered with commercial aircraft transmissions. For this same reason, bugs seldom operate in the standard a-m broadcast band.

Other factors influencing the choice of an operating frequency are the required length of the antenna and the availability of commercial receivers for monitoring. The requirement that the antenna be as short as possible so that it can be concealed dictates the use of high frequencies. The availability of economically priced sensitive receivers has led to the use of many bugs either in or close to the standard fm band of 88 to 108 megahertz, or just outside this band. Police departments use the 30- to 50-megahertz bands and the 150- to 174-megahertz bands, making this an inadvisable frequency range for illegal transmitters. In actual practice, almost all bugs operate somewhere between 60 and 112 megahertz. Usually, commercially available fm receivers can be modified to cover a frequency in this range with a minimum of trouble. The use of fm helps greatly to combat radio noise.

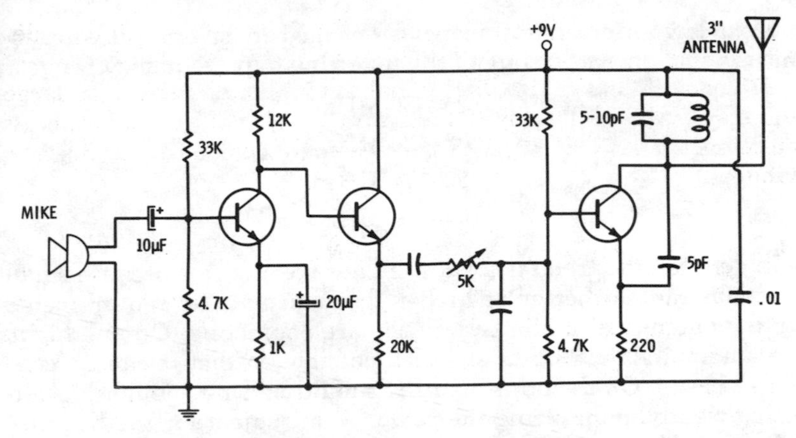

Fig. 12-4. Circuit for an fm bug.

Fig. 12-4 shows the circuit diagram of a bug that is commonly used. It consists simply of an emitter-coupled oscillator that is frequency-modulated by a single audio amplifier. More elaborate devices use more stages, but the basic principle is the same. These devices can be made very small. Several commercially available units are shown in Fig. 12-5.

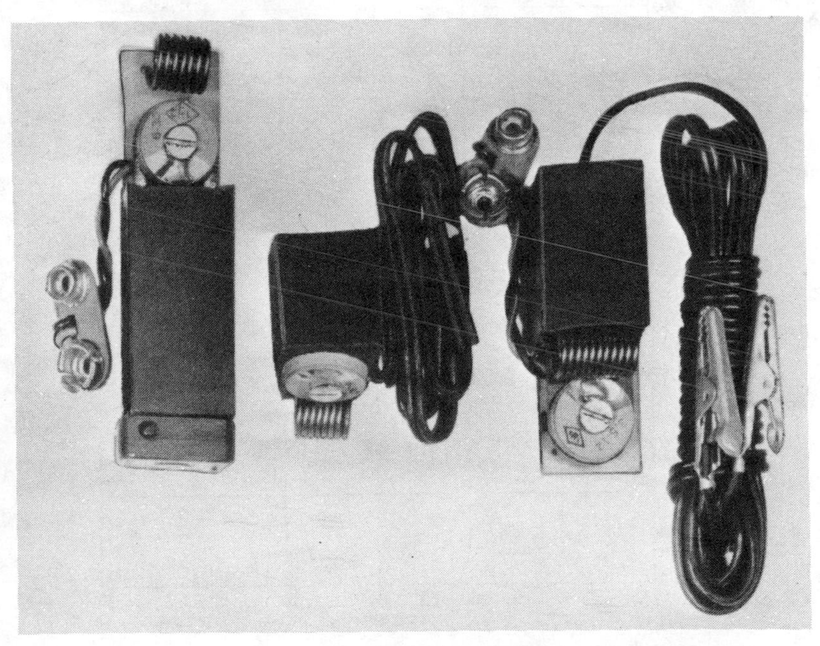

Fig. 12-5. Commercially available bugging transmitters.

Most law enforcement agencies engaged in bugging use transmitters that operate in one of the following four frequency ranges:

25– 50 MHz
88–120 MHz
150–174 MHz
400–570 MHz

In general, the trend is toward higher frequencies because small antennas radiate better at higher frequencies. The requirements for the antenna of a bugging device are conflicting. On one hand, the antenna should be as short as possible so that it can be concealed easily. On the other hand, it should be long enough to radiate efficiently, minimizing the power requirement for the bug.

In general, an antenna shorter than about a quarter wavelength will not radiate efficiently because its losses will be high. The higher the frequency, the shorter the wavelength, so the easier it is to build an efficient antenna. Fig. 12-6 shows a plot of the length of a quarter wavelength as a function of frequency. Note that at 30 MHz a quarter-wave antenna would be nearly 250 centimeters

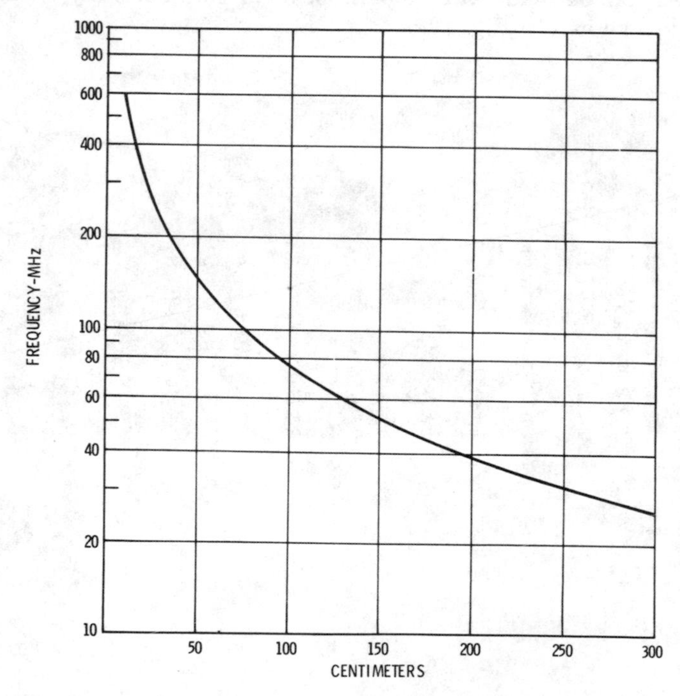

Fig. 12-6. Length of quarter wavelength for various frequencies.

(8.2 feet) long, whereas at 500 MHz it would be only about 15 centimeters (5.9 inches) long. This explains the trend toward higher frequencies.

Another consideration in the selection of a frequency for operation of a bug is interference. Bugs used by police sometimes operate on police frequencies under a regular license. In such a case, the frequency can be kept clear of interference. Surreptitious bugs must be used on any frequency the eavesdropper can use without either detection or interference.

Operating Range

In general, the signal from a bug should be receivable at a distance great enough that the eavesdropper can remain out of sight and in a location where he will not easily be detected. This is true whether the eavesdropper is a law enforcement officer or a criminal. The range that can be obtained by using a small bug depends on many factors. One such factor is the power output. This means the actual power output at the fundamental frequency. Since bugs must necessarily be small, the transmitter often is merely a modulated oscillator.

Most modulated oscillators have a very high harmonic content in their outputs. A broad-band power measurement that measures both the fundamental frequency and the harmonics can be very deceptive. It is only the power at the fundamental frequency that is effective.

The effect of signal attenuation on range is plotted in Fig. 12-7. In this plot, the point 1.0 on the axis corresponds to the maximum range, whatever it may happen to be. From the plot, it can be seen

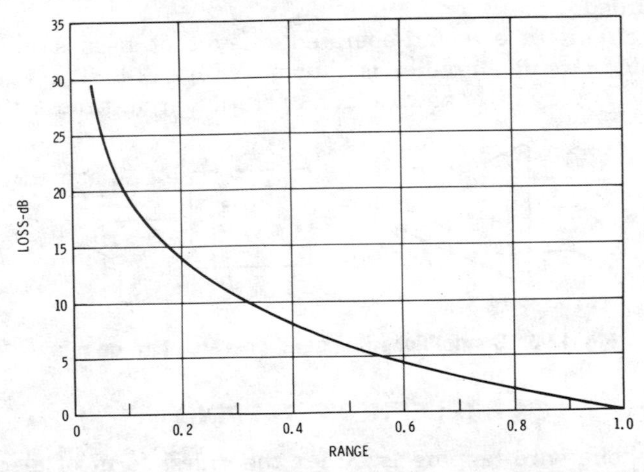

Fig. 12-7. Normalized range as a function of signal attenuation.

that 6 dB of attenuation will cut the range in half and 20 dB of attenuation will cut the range to 1/10. Attenuation may be introduced by buildings, structures, etc. Thus, it can be seen that the range obtained from a given bug will vary drastically from one situation to another.

Sound-Operated Relay

One of the disadvantages of electronic eavesdropping devices is that their life is limited by the small batteries usually used with them. When operated continuously, bugging devices have a fairly short life. Continuous operation of a tape recorder at a monitoring point is equally wasteful. If a recorder is operated so that it continuously monitors an area, it may run for 24 hours and record only a few hours, or even a few minutes, of conversation.

This limitation may be overcome by the use of a sound-operated relay. This is a small device having a very low power drain that will actuate a relay only when sound is picked up by a microphone connected to it. When it is connected to a transmitter, the transmitter will normally be turned off. When a sound is picked up in the monitored area, the relay will turn on the transmitter. A time-delay circuit will keep it on until a few seconds after the sound has died out. This has the dual advantages of conserving battery life and keeping the transmitter on the air only when it is transmitting something. This greatly reduces the chances of its being accidentally discovered by someone idly tuning through the frequency of operation.

A sound-operated relay can also be used at the monitoring point so that a tape recorder will run only when there is something to be recorded.

The circuit of a sound-operated relay that uses a miniature integrated-circuit amplifier is shown in Fig. 12-8. The amplifier, which uses very little power, is in operation at all times.

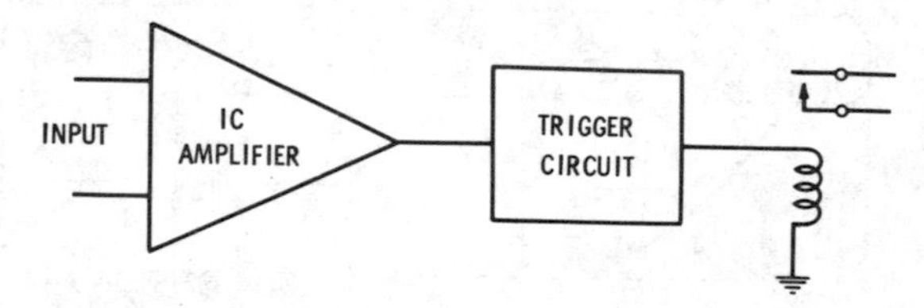

Fig. 12-8. IC amplifier with sound-operated turn-on relay.

TELEPHONE TAPPING

Telephone wire tapping is by far the oldest form of electronic eavesdropping. In its simplest form, the telephone tap consists of simply bridging a headset or a tape recorder across the line. Usually, a high-impedance amplifier is used to prevent disturbance of the telephone circuits. In recent years, telephone bugging has become much more sophisticated. Miniature fm transmitters are available that can be concealed within the telephone set to broadcast not only telephone calls but also any conversation that can be picked up by the telephone transmitter. Fig. 12-9 shows a complete fm transmitter circuit which can be constructed around the carbon microphone element of a telephone. This unit can be installed in a telephone in a few seconds. The cover need only be unscrewed from the phone and the regular microphone element be replaced by the new one containing the transmitter.

A very elaborate bug that uses the telephone is shown in the block diagram in Fig. 12-10. The advantage of this unit is that after a phone number is dialed, the line is connected for a short time before the bell begins to ring.

When the unit of Fig. 12-10 is installed in a user's phone, the eavesdropper may listen to conversations in the room from any location where he can get telephone service. He first dials the number of the bugged phone; then before the bell starts to ring, he puts a 500-hertz signal on the line, usually by blowing a small whistle. The 500-hertz signal passes through the 500-hertz filter

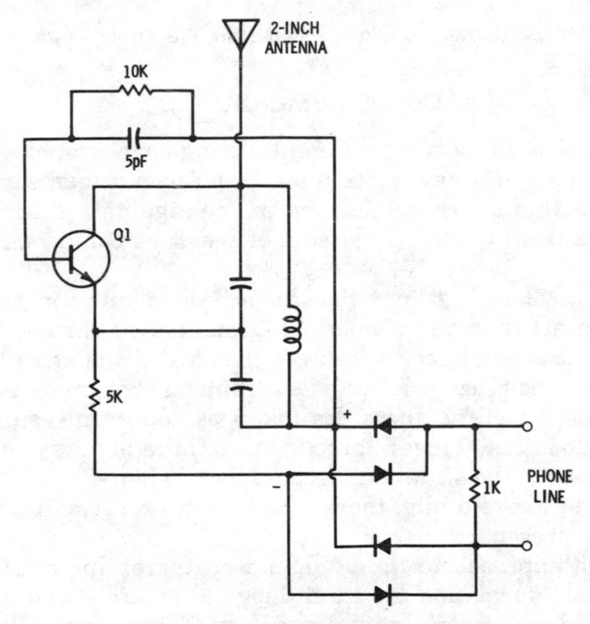

Fig. 12-9. Circuit of an fm transmitter which can be built into a telephone transmitter button.

in the bug and operates a sound-operated relay similar to the one described earlier in this chapter. This relay effectively answers the phone and disconnects the ringing circuit. The bugged phone is now connected to the eavesdropper's line. He can listen to any conversations that reach the telephone transmitter with no one in the room being aware of it. If anyone else attempts to dial the same number while the bug is in operation, he will get a busy signal. If the owner of the bugged phone wants to make a call, the eavesdropper can hang up and the line will be restored to normal. This arrangement can be used to bug a room through a telephone, but cannot be used to eavesdrop on telephone conversations.

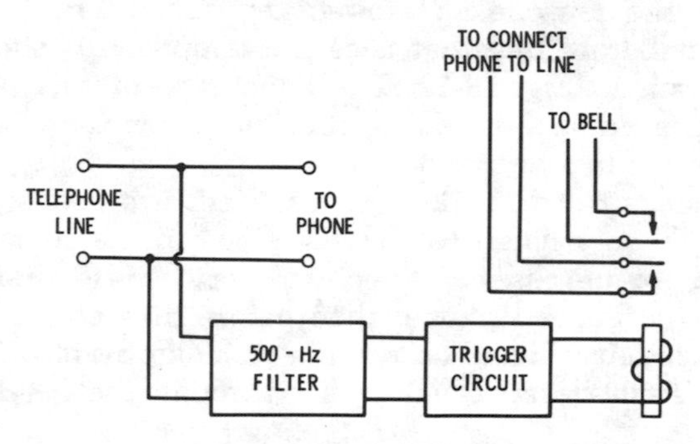

Fig. 12-10. An elaborate bugging system using the victim's own telephone.

COUNTERMEASURES

The problem of locating and neutralizing eavesdropping devices has given rise to a new profession. Several management and security consultants are specializing in "debugging" and in providing secure areas for the discussion of sensitive company information.

In general, locating bugs that have been planted in an area is more of an art than a science. As was mentioned earlier, the state of the art at present seems to favor the eavesdropper rather than the law-abiding citizen. When eavesdropping is strongly suspected in a business facility, the steps taken as countermeasures often assume "cloak and dagger" proportions. There is always some uncertainty about the situation, because even when all possible measures fail to locate a bug, there is still no absolute assurance that one is not present.

The first approach to providing a secure area for discussion of proprietary information is a careful physical search. The suspected area should be carefully inspected, and every possible hiding place examined carefully. Particular attention should be paid to new additions to the area such as pictures, plants, and lamps. These are favorite hiding places.

Probably one of the most successful ways of planting a bug in a business establishment is for a visitor to simply leave something behind. Typical items are briefcases, hats, raincoats, and catalogs. Any of these items might well conceal a miniature transmitter.

SEARCH METHODS

Wired microphones can often be located by first discovering the wires. A metal locator (see Chapter 8) is also helpful in locating concealed items.

In addition to conducting a physical search, most debugging experts believe in making rf field-strength measurements. These are usually made at the frequencies that are commonly used, but to be perfectly safe would require making measurements at all possible frequencies. This is usually considered impractical for normal business searches.

Many field-strength meters designed specifically as "bug locators" are commercially available. Fig. 12-11 shows the circuit diagram of a simple field-strength meter that will locate just about any hidden transmitter if it is brought close enough. It consists simply of a broad-band diode detector and a high-gain, integrated-circuit amplifier. The meter will deflect on very weak signals, and in some locations the sensitivity must be turned down to avoid indications caused by local broadcast stations. The signal causing the deflection may be identified by listening to the signal. A small speaker plugged into the phone jack will help in locating bugs. The gain of the amplifier is so high that when it is brought close to the microphone of the bug, feedback will occur and the device

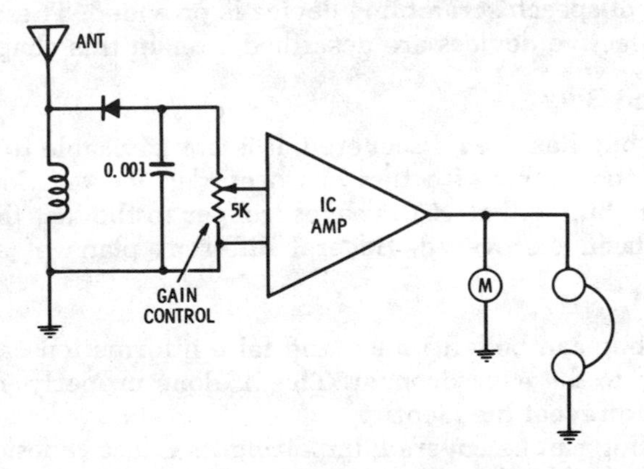

Fig. 12-11. High-gain, broad-band, field-strength meter.

will howl. When trying to locate bugs with sound-operated relays, the searcher will usually talk continuously while conducting his search. If he hears his own voice in the headphones, it is positive indication that a bug is present and is being triggered by his voice.

Other commercially available field-strength meters are tunable. These usually have a higher sensitivity but require much more time to operate.

Protection of Business Secrets

Even though an area has been thoroughly searched and found to be clear of eavesdropping devices, it is still wise to treat business secrets at least as carefully as any other property of the same value would be treated. It is rather ironic that some firms will go to great lengths to provide proper protection for their material assets, but will freely discuss proprietary business secrets worth thousands of dollars in completely insecure areas such as restaurants, bars, and motel rooms. Whenever it is suspected that eavesdropping devices might be in use, conversations should be held in subdued tones and background noise (such as a radio playing in the background) should be provided.

Telephone Inspection

Since telephones are frequently tapped and also used as hiding places for bugs, they should be thoroughly searched. Usually, this is best accomplished with the assistance of a representative of the telephone company. Being familiar with the equipment, he can quickly spot signs of tampering or the presence of alien equipment. Because of the many different places that a telephone line may be tapped, telephones should never be considered as secure channels of communication for highly sensitive information unless some sort of speech scrambling device is provided. These interesting and effective devices are described later in this chapter.

Discovered Bugs

After a bug has been discovered, it is not advisable to touch the bug until the course of action has been decided on. Touching or moving the bug will alert the eavesdropper to the fact that his device has been discovered. Several different plans of action are possible:

1. The bug can be left intact, and false information can thus be given to the eavesdropper. This, if done properly, may force him to reveal his identity.
2. The bug can be covered, impairing its effectiveness; then the area can be placed under surveillance. The eavesdropper,

finding that his bug is not operating properly, may return to pair or replace it.

3. The bug simply may be removed. This is apt to create a false sense of security, however, because two bugs are often used in the expectation that one will be found and the victim will feel that he is now safe.
4. The most advisable course of action in most cases is to leave the bug intact and report the matter to the police and an attorney.

ILLEGAL JAMMERS

Although transmitters that are designed to interfere with transmission of radio signals are forbidden by law, they are widely used and should be discussed. In fact, it seems that many business executives who have little electronic knowledge feel very secure when they know that any possible radio transmission from a conference room will be jammed.

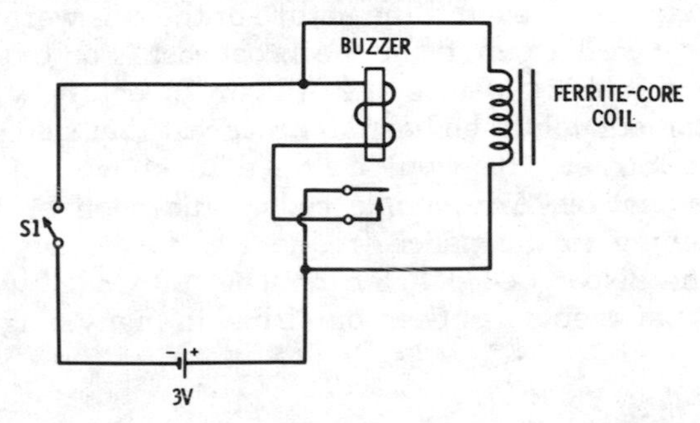

Fig. 12-12. A buzzer-type jammer (illegal to use).

Commercially available jammers vary considerably in their construction. The purpose of all these devices is to radiate broad-band noise that will mask the signal from a bug. While doing this, they usually do just as well at jamming radio and television signals in the area.

A schematic diagram of a buzzer-type jammer is shown in Fig. 12-12. Some of these units are designed to fit into a fountain-pen case so that they can be carried about.

SECURE COMMUNICATIONS

One of the most serious eavesdropping problems today is the monitoring of police radio broadcasts by burglars hoping to avoid apprehension. More and more often, police answer a burglar alarm only to find that when they get there the burglar has already gone. Portable receivers that will monitor police frequencies are available at economical prices and can be purchased by anyone. Most of these receivers are used by law-abiding hobbyists who follow the police in their work and do no harm. But their availability makes it possible for a burglar to carry a receiver with him on a burglary and leave immediately if he hears a police cruiser being dispatched to the area where he is working. Some cities have passed ordinances forbidding the use of portable radios tuned to police frequencies, but these ordinances are usually ineffective.

Coded Police Messages

The most obvious approach to the problem is to use coded messages for dispatching police cars. This has been done for many years but has had almost no effect. First of all, the code must be very simple and easily remembered if all police officers are to use it and react to it instantly without recourse to a code book. It would be practically impossible to have a code name for each street in a city. Thus, even if the nature of the call were coded, the burglar could still profit from the broadcast. For example, if a burglar were robbing a store at 210 Main Street, he would leave the scene immediately if he heard a broadcast that said, "Signal 50 at 210 Main Street"; he wouldn't need to know just what was meant by signal 50. Any simple code could soon be broken by merely listening to the police frequencies for a few weeks and following the dispatched police cars. The "10 Code" widely used by police departments has been published in many magazines and books.

Military Services

For years the military services have had very elaborate secure-communications systems for fixed locations, but even they have had trouble transmitting coded messages in the field where a quick response is required with no time for elaborate decoding. During World War II, the British forces used Irish troops who spoke the little known Gaelic language for secure communications. The United States forces have used American Indians in the same manner.

Speech Scramblers

The electronic answer to the problem of providing secure communications is the use of speech scramblers. The scrambler is a small device that is inserted between the microphone and the transmitter. It has the effect of making the broadcast unintelligible unless a receiver incorporating the proper descrambler is used.

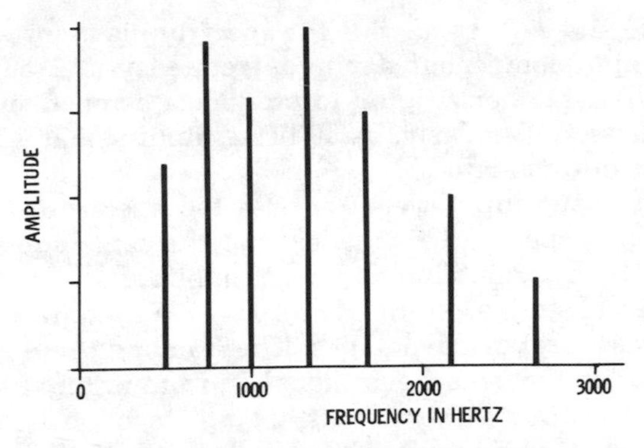

Fig. 12-13. Frequency spectrum of speech sounds.

Some very elaborate electronic coding systems are used by military agencies, but these are usually classified "SECRET" and are far too elaborate for regular police or business use. The speech scramblers that are in common use today operate on two basic principles—inversion and splitting.

Normal speech sounds consist of many different frequency components, each having a different amplitude. A plot of amplitude versus frequency of a typical speech sound is shown in Fig. 12-13. In the inversion scrambler, the order of these components is reversed. The result is that when the signal is picked up on a regular receiver, the signal sounds like monkey chatter—something like a single-sideband signal that is not tuned in properly.

Fig. 12-14 shows a block diagram of a simple inverter. The input signal consists of speech sounds having components in the frequency range of 250 to 2750 hertz. These signals are fed to a modulator where they are heterodyned with a signal from a 3-kilohertz oscillator. Two different sets of signals are produced in the modulator—the sum of the speech frequencies and the 3-kilohertz signal, and the difference between the 3-kilohertz signal and the speech frequencies. A low-pass filter in the output lets only the difference frequencies pass. Thus, the output frequencies are

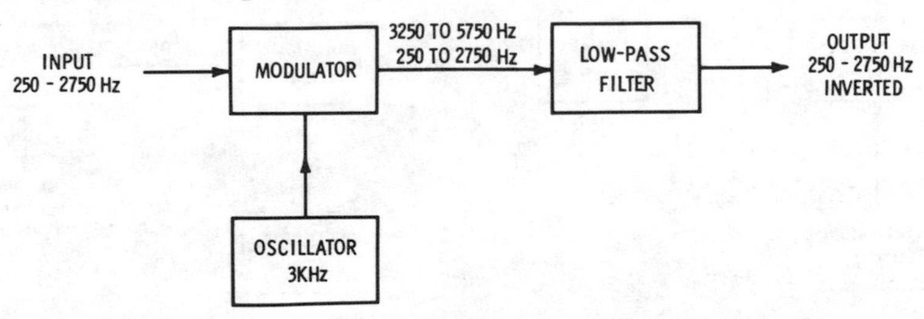

Fig. 12-14. Block diagram of a simple speech inverter.

between 250 and 2750 hertz, but the spectrum is inverted. For example, an input component having a frequency of 2750 hertz will beat with the 3-kilohertz signal to produce a component of 3000 − 2750, or 250 hertz. Similarly, a 250-hertz input signal will produce a 2750-hertz output signal.

Note that if the input were an inverted spectrum, the output would be plain speech. This means that the same equipment can be used for both scrambling and descrambling.

Although the simple circuit of Fig. 12-14 is used in some scramblers, it has several disadvantages. Chief among these is that it is difficult to filter the 3-kilohertz signal and the original speech signal from the output. Adjustments are apt to be critical, and the unit is hard to keep properly adjusted in the field. If even a small amount of normal speech filters through, it will serve as a clue to the listener and help him to at least guess the mesage.

An arrangement that produces the same result but is much easier to adjust is shown in Fig. 12-15. This system uses double modulation. In the first modulator, the speech is heterodyned with a high-frequency signal, for example, 13,000 hertz. Only the high-frequency components are passed on to the second modulator, which operates exactly 3000 hertz higher in frequency than the first modulator. Here, a filter selects the low-frequency, or difference, components. In this arrangement, if the input signal has a frequency of 2750 hertz, the output of the first modulator will be a frequency of 15,750 hertz. The output of the second modulator is then 16,000 − 15,750, or 250 hertz. Thus, the arrangement produces exactly the same result as the simpler circuit of Fig. 12-12, but, with this system, adjustment and maintenance are much more straightforward.

The chief disadvantage of the simple inverters just described is that their signal can be descrambled by using an ordinary signal generator, together with a regular receiver. If, using the frequencies in the above examples, a signal generator is tuned exactly

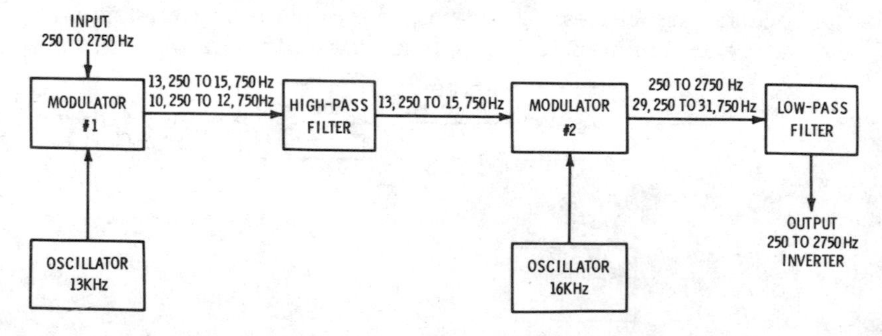

Fig. 12-15. A double-conversion speech inverter.

3000 hertz below the carrier frequency and fed to an ordinary receiver, together with the inverted signal, the output will contain plain speech. There will be an annoying 3000-hertz beat present also, but the signal will be reasonably intelligible.

In spite of this limitation, a simple inverter can be quite effective. The first consideration in using a scrambler is to use it only where necessary. For example, ordinary police calls can be handled with plain speech, and scrambler speech can be used only for dispatching cars to an area where a crime is in progress. In this way, there will be less opportunity for anyone to record the scrambled signals and develop a descrambler.

More elaborate scramblers use a principle called *splitting,* in addition to inversion. In these systems, the speech spectrum is broken into several separate bands by filters. Each of these bands may be transmitted plain or inverted, and the bands themselves may be interchanged. In transmission of transoceanic telephone conversations, the speech spectrum of 250 to 3000 hertz is split into five bands, each 550 hertz wide. The bands are interchanged and may be transmitted as either plain speech or inverted. The arrangement is changed once every twenty seconds.

Fig. 12-16 shows a block diagram of a system that not only will invert speech, but also will split the speech spectrum into two halves that can be interchanged. Although the entire speech spectrum from 0 to 3000 hertz is applied to both channels, the filters in

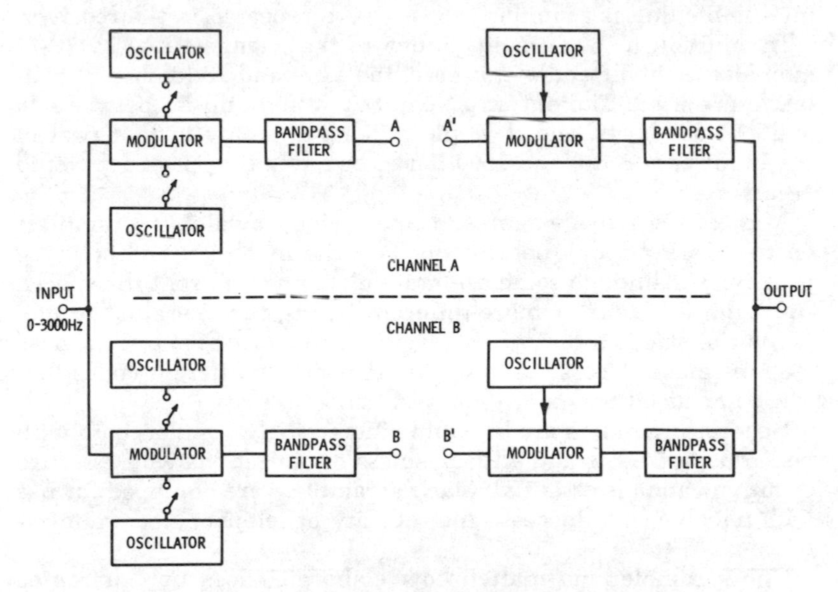

Fig. 12-16. Two-channel splitter and inverter for speech signals.

the system restrict the effect of channel A to those components from 0 to 1500 hertz, and the effect of channel B to those components between 1500 and 3000 hertz.

First, consider how the system splits the speech spectrum without inverting it. In this case, the 10- and 8.5-kilohertz oscillators are used. The entire speech spectrum, 0 to 3000 hertz, is heterodyned with a 10-kilohertz signal in channel A. The output of the modulator contains both sum and difference components. It has components in the bands of 7 to 10 and 10 to 13 kilohertz. Only those components between 10 and 11.5 kilohertz can get through the bandpass filter. Thus, the output at point A is due only to those components of the input lying between 0 and 1500 hertz. Channel B operates in the same way; its output is in the same frequency range but is due to the components of the input lying between 1500 and 3000 hertz. Neither of the bands is inverted.

If in Fig. 12-16, point A were connected to point A′, and point B were connected to point B′, the heterodyne process would be reversed in the second half of the system, and the output would be just like the input. If, on the other hand, point A were connected to B′, and point B were connected to A′, the two halves of the spectrum would be changed. That is, the output of channel A, due to the 0- to 1500-hertz portion of the input would be shifted to lie between 1500 and 3000 hertz. Thus, the two halves of the speech spectrum would be interchanged. Such a signal would not be intelligible unless a similar process were repeated in the receiver.

In addition to being split, either of the channels in Fig. 12-16 may also be inverted. In this case, the 11.5- and 13-kilohertz oscillators are used. The output at point A would still lie between 10 and 11.5 kilohertz, and it would still be due only to that part of the input between 0 and 1500 hertz, but now the spectrum would be inverted.

Most of the more elaborate commercially available scramblers on the market today operate on both the inversion and splitting principles. Although some are more elaborate and split the speech spectrum into four or more different bands, the operating principle is the same as that of the system shown in Fig. 12-16. These systems make it possible to use a great many different codes that can be changed frequently for additional security.

Speech scramblers are becoming increasingly popular, both with police departments and with business organizations where secure communication is essential. Many scramblers are equipped for use with telephones to increase the security of telephone communications.

The scrambler, no matter how elaborate, does not guarantee permanent security of communications. It is always possible to

tape the scrambled message and eventually decode it by trial and error. The chief advantage of scrambling is that, with a good scrambler, it takes a great deal of time to break the code. This is long enough to protect emergency communications that require immediate action; by the time the opposition has decoded them, they will already have been acted upon and the need for secrecy will be past.

COMPUTER SECURITY

One of the most serious security problems facing both industry and government today is the security of computers. The number of crimes that can be committed in connection with a modern computer staggers the imagination. Personal records of all types are kept in computer memories, and the possibilities for invasions of privacy are almost unlimited. Computers are used to keep all sorts of company records, and these records may be stolen or destroyed. Increasingly, computers are used to handle funds, which may be diverted to the wrong account by a simple program change. If the computer is to fulfill its promise, a means must be found to make it more secure.

There are several ways in which unauthorized access to a computer may be gained. Some of these include:

1. *Unauthorized access to a regular terminal.* Someone may enter the premises after hours, much as a burglar would, and merely perform his criminal activities at a computer terminal. This type of crime is insidious in that, unlike a burglary, there is no way of even knowing the crime has been committed until it is much too late.
2. *Unauthorized access through a remote terminal.* Many computers can be accessed by any number of terminals that are far removed from the main facility and the regular security people. From such a vantage point, the perpetrator of the crime can leisurely try various approaches, often with the aid of the computer itself.
3. *Access through wiretapping.* Many computers transmit data from one point to another either through private dedicated lines or through the telephone system. By means of the same wiretapping methods used to eavesdrop on telephone conversations, data may be intercepted or false data may be entered into the system. Wiretapping of data is often more serious than eavesdropping on conversations because data is of a permanent or semipermanent nature and the eavesdropper has plenty of time to execute his acts.

4. *Access through microwave links.* Many data systems include microwave links that transmit data through narrow radio beams. These radio paths often pass over relatively uncongested areas where an eavesdropper can raise an antenna to intercept them without detection.
5. *Access through electromagnetic eavesdropping.* Computers use sharp waveforms in their operation, which means that there is usually some radiation from them. It is possible to record the radiation from a computer and later decode the data with the aid of another computer. Even though the radiation from a computer can usually be picked up only a short distance away, it is possible to plant a bug in the facility that will relay the signal over a much greater distance.
6. *Access through radio links.* Many law enforcement agencies use computer terminals installed in vehicles; these terminals interface the computer through a radio link. Eavesdropping on these transmissions is merely a matter of recording the signals and later decoding them with the aid of a small computer. Once the code is broken, the messages can be intercepted and typed in plain language. With this same arrangement, it may be possible for a criminal to access such a computer through an ordinary mobile radio and enter false data into the system. This could range from giving false information about an enemy, to reporting that a car he had stolen has been recovered so that police would no longer be looking for it.

There are well over a million computer experts in the country today, any one of whom might be interested in illegal access to a computer. The computer security problem is aggravated further by the fact that foiling a computer is a challenge, much like winning a chess game. As such, the activity attracts many who would not ordinarily engage in criminal activities but who are fascinated by the challenge.

Risk Evaluation

The first step in combating computer crime is to recognize that the problem exists and to have some idea of the magnitude of the risk. This means assigning some value to all the data kept in the computer memories. As with all other types of crime, it can be assumed that a criminal will go to any length if he feels that the reward is commensurate with the risk. In connection with computer crime, there are both extremes. The reward may be very great, or the criminal may feel that the risk of being caught is very slight. Some computer criminals have not been prosecuted be-

cause the victim has thought that the publicity incidental to prosecution would disclose a system weakness that would be exploited by others.

The types of risks involved in a computer system vary with the application. They include, but are not necessarily limited to, the following:

1. *Loss or destruction of data.* This type of act is usually perpetrated by one who is vindictive or is championing a radical cause. The reward to the criminal is questionable, but the loss to the victim may be very severe. One of the best defenses is to keep duplicate files at a remote location.
2. *Diversion of funds or data.* Any computer that handles funds or valuable data is fair game to the computer criminal. The rewards can be very great, and the loss to the victim equally great. Diversion of data is harder to detect than diversion of funds. If funds are transferred, they will eventually be discovered as missing from the proper account. Data, on the other hand, can be stolen without being destroyed. A competitor can obtain a list of major customers together with figures on their volume of business, and there will be no way of knowing that the data has been stolen. The original files will still be intact.

Once the risk has been carefully evaluated, a decision can be made on the extent of the security measures to be taken. In general, computer security is expensive, and some experts feel that because of the nature of the problem, absolute security is impossible. As with other security situations, measures can and should be taken to minimize risks.

Facility Security

The first step in computer security is to provide security for all facilities where access to a computer can be gained. This starts with secure doors and locks and a good intrusion alarm that will deter unauthorized entrance to the facilities.

In general, a means should be provided for recording any unauthorized entry. If computer crime can't be entirely prevented, the next best step is to alert the proper personnel that one may have taken place. If a computer facility has been subject to an intrusion, a computer audit may be performed to determine whether or not data has been altered.

Computer Audits

A good security check on a computer facility is an audit program that will determine how much data has been manipulated

over a period of time. In many facilities, the normal fluctuation of accounts can be predicted with reasonable accuracy. If unusual swings in data are found in the audit, espionage can be suspected.

Electronic Security Systems

Many electronic measures can be taken to increase the security of a computer facility. In addition to the previously mentioned facility security, access control measures can be taken. These can be any of the systems described in Chapter 10 or the more sophisticated electronic identification methods covered in Chapter 14. In any computer system where great risks are involved, only authorized persons should have access to the computer facilities and terminals.

Fig. 12-17 shows a block diagram of a system that will keep a record of who accesses a computer and at what time. Each terminal has a means of identifying the operator. This can be a separate key switch for each authorized operator, magnetic access-control cards, or in cases of extremely high risk, fingerprint identifiers.

The access record is kept on magnetic tape that cannot be reversed. Thus, the computer can't be programmed to erase the access record.

This record will not prevent crime, but it will act as a deterrent. Furthermore, it may help in apprehending a criminal after a crime has occurred.

One way in which electronics can be used to increase the security of a computer system is with cryptography; that is, the encoding of the data into a secret code. The science of cryptography has been highly developed by military services for transmissions of vital national importance. Unfortunately, there are not many expert cryptanalysts in industry. There are more apt to be cryptanalysts with military experience attempting to break a code than to compile codes for security purposes.

The encoding of data is similar to encoding a message in English. It consists basically of substituting a code character for each character in the message.

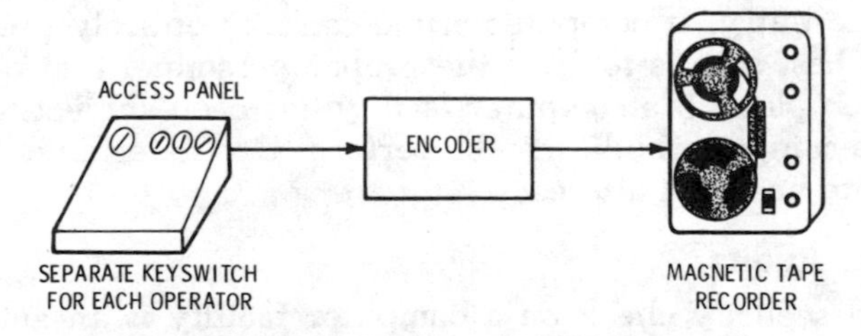

Fig. 12-17. A system for recording computer access.

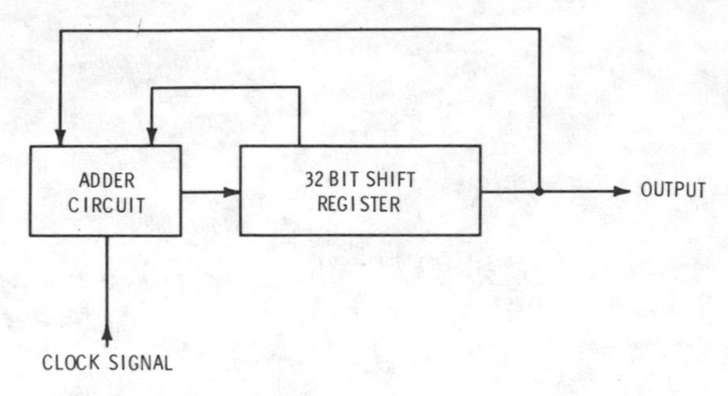

Fig. 12-18. A circuit that will develop a nearly random digital sequence.

To one unfamiliar with cryptography, most codes appear to be very secure, but this is deceiving. To the expert, breaking a code is not nearly as complicated as it seems. Fig. 12-18 shows a system that will generate a long series of binary 1's and 0's. It consists of a series of shift registers with feedback from two points along the string. This circuit can develop a series of almost a billion 1's and 0's that appear to be nearly random. In the system, this string of digits is added to the data at the transmitting end and subtracted at the receiving end. The device itself can be constructed on a printed-circuit card, or even on a monolithic integrated circuit. The operator need not have any idea of what is in the encoder or how it works.

To the inexperienced, this encoding with a string of nearly a billion binary digits would appear to provide absolute security. It doesn't, however, for two reasons. First of all, anyone stealing data knows something about it. No one would go to the trouble of stealing anything that was of no value to him. This advance knowledge gives him some idea of what is in the data. It might not be much, but it is enough for an experienced cryptanalyst to decode a message with the aid of another computer.

Breaking a code usually boils down to solving a number of linear equations. If not much advance information is available, the number of equations that must be solved is large, but if another computer is available, this is no problem.

More sophisticated encoding techniques are available using nonlinear circuit elements. These codes are much harder to break.

CHAPTER **13**

Automobile Protection

One of the most common types of thefts in the country today is car theft. Cars are stolen for resale, for the value of the parts and accessories, and for use in committing other crimes. It is unfortunate that so few cars are equipped with electronic security systems because it is easy to provide a lot of protection at a reasonable cost.

One of the reasons why the automobile is so vulnerable to theft is that it not only provides a built-in means of escape for the thief, but also deprives the owner of a means of pursuit.

When considering methods of protecting an automobile from theft, consideration must be given to the fact that most car thefts are accomplished by professionals. Ordinary burglars run the gamut from bungling amateurs to skilled professionals, but car thieves seem to number more highly skilled operators in their ranks. The reason is probably that one car is very similar to another, and a skill learned on one type of car can readily be adapted to another.

At any rate, a professional car thief can gain access to a car, hot wire it, and drive away in less than a minute. In fact, much of his success depends on executing the crime rapidly. This gives a clue to the type of protection that will be most effective, namely, anything that will increase the amount of time required to steal the car.

The risk of getting caught stealing a car increases drastically with the time required to execute the theft. Police departments have very efficient reporting systems for tracking stolen cars. As soon as a car theft is reported, the information is introduced into

a nation-wide computer network. Once the information is in the computer, it is just a matter of time before the car will be recognized as stolen. Thus, the car thief needs to gain as much time as possible before the theft of the car is discovered. If stealing a particular car proves to be a time-consuming or unfamiliar process, he may readily abandon the job and look for one that is easier to steal.

The second deterrent, after making the job difficult, is to provide an attention-getting alarm that is difficult to foil. The last thing that a thief wants is attention.

IGNITION PROTECTION

The problem with most proposed ignition-protection systems is that they can be "jumped" rather easily by a professional. The best protection against jumping is to make the situation unfamiliar. Fig. 13-1 shows a deceptively simple, but quite effective, system of ignition protection. A concealed wire is run from where the ignition hot lead attaches to the coil through a concealed switch to the automobile ground. The hot wire is broken at some inconspicuous place, and a fuse is inserted. The fuse is rated high enough to handle the normal ignition current.

When the car is parked, the hidden switch is closed, thus connecting the hot wire of the ignition to ground. If the ignition is turned on, or the ignition key is jumped, the fuse will blow, interrupting the ignition current. The thief will not realize that a fuse has blown but will think either that he has tried to steal a car that won't start, or that there is an unfamiliar situation which he is not competent to handle.

If the thief attempts to connect a hot wire directly to the terminal of the coil, he will be connecting a wire between the battery supply and ground, and the wire will become red hot very quickly. This alone is often enough to discourage even a professional.

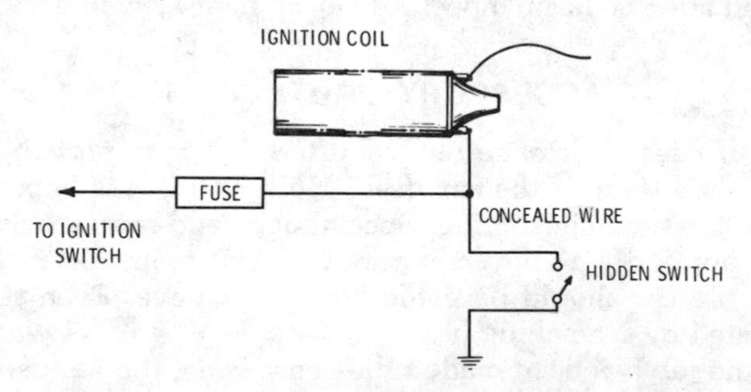

Fig. 13-1. A simple but effective ignition-protection system.

HOOD AND TRUNK LOCKS

Probably the most effective step that can be taken to prevent car theft is to install good hood and trunk locks, such as those shown in Fig. 13-2. The ordinary trunk lock can be punched out so quickly that it is usually quicker to open a trunk in this way than to use a key. Most hoods have no protection. The use of a hood lock is inconvenient when checking the oil level, but the amount of protection it provides is worth the inconvenience.

Any lock, even a heavy one with a strong chain, can be broken, but it will take time. As a further deterrent, both the hood and trunk should have tamper switches that will initiate an alarm when an attempt is made to open them.

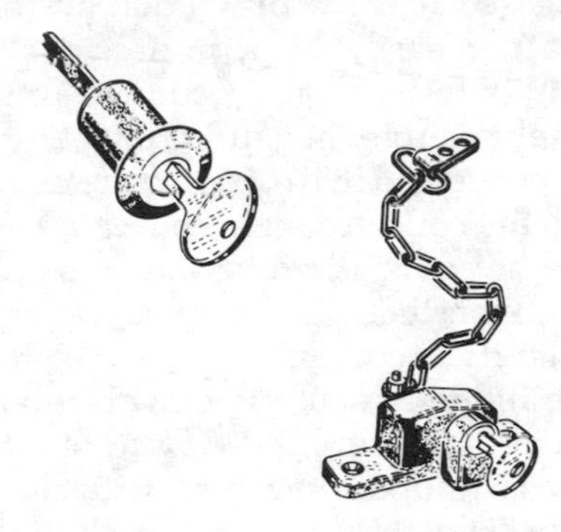

Fig. 13-2. A typical hood or trunk lock.

The usual approach to stealing a car that might be equipped with an intrusion alarm is to open the hood and disconnect the battery cable in less than a minute. The alarm will of course be triggered, but it will stop as soon as the battery cable is disconnected. The thief can then find the alarm wires and disable the alarm. He can jump the ignition and drive away with impunity. If he finds that it will take more than a few seconds to get the hood open after he has tripped the alarm, he may well give up.

ACCESSORY PROTECTION

Theft of tires, radios, stereos, and CB radios is probably more common than theft of the car itself. Here again, it is important to make the job time-consuming. Special bolts and nuts are available that are not easily removed without special tools. Everything of value in the car should be made hard to remove. Even tires can be protected by special lug nuts.

Once the job has been made time-consuming, the next step is to provide a good alarm. If all of the alarm circuitry is mounted un-

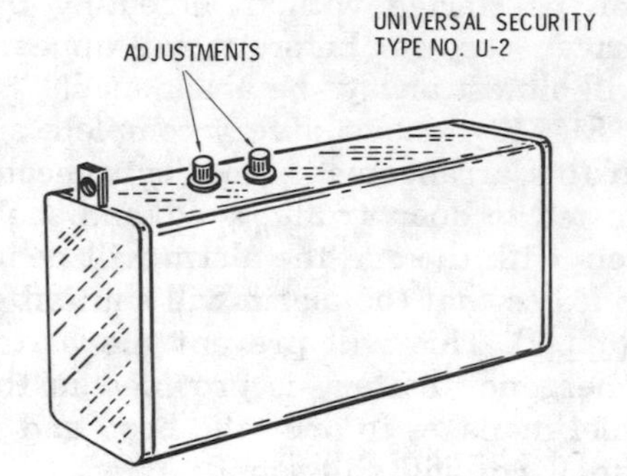

Fig. 13-3. Automobile shaker alarm switch.

der the locked hood, there is less likelihood that it can be foiled. With proper alarms, door switches are usually adequate to protect accessories inside the car.

To protect the tires, an alarm is needed that will trip when the car is vibrated, such as when it is jacked up. The shaker switch shown in Fig. 13-3 is made specifically for this purpose. It will trip the alarm as soon as the car is vibrated or moved. Naturally the shaker switch should be mounted where it cannot easily be detected until it is too late.

COMPLETE PROTECTION

The measures described in the preceding paragraphs are usually sufficient to provide a great deal of protection for an automobile. They cannot totally prevent theft, but, by making it difficult, they will make it time-consuming and will therefore constitute an effective deterrent.

Where it is thought necessary, additional measures may be taken. A feature that is commercially available but not commonly used is a solenoid valve in the fuel line. The solenoid valve is controlled by a hidden switch or electric combination lock. The car

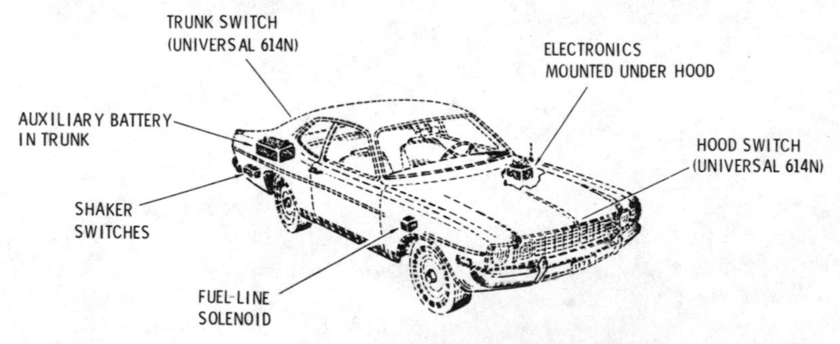

Fig. 13-4. A complete automobile-protection system.

can be started without defeating the system, but it cannot be driven very far before it will appear to have run out of gas and will almost always be abandoned.

Fig. 13-4 shows a very complete automobile protection system. In this arrangement, an electromechanical alarm system is connected to door, trunk, hood, and shaker switches. If anyone tampers with the car, the alarm will be initiated. A time delay is provided so that the alarm will shut off soon after the tampering has stopped. This will prevent the battery from running down. An emergency battery is provided in the trunk so that even if the thief manages to open the hood and disconnect the battery cable, the alarm will still sound.

A fuel-line protection system is installed so that even if the thief detects all of the other protective measures, he will soon think he has run out of gas.

CHAPTER 14

Personnel Identification

At one time, identification of personnel was considered to be strictly a human function. Security guards and bank tellers were charged with the responsibility of identifying the people they dealt with. The cost of using a human being to recognize and identify other people is high and is not economically feasible in all situations. It has been estimated that it takes 5.2 security guards to adequately cover one entrance 24 hours a day when allowances are made for a 5-day week and the normal vacation and lost time.

Human guards are not infallible. They can make mistakes or in some cases collaborate with an intruder. A system is needed that will automatically identify authorized persons, or at least assist the guards who are responsible for the identification.

Automatic equipment, such as the automatic tellers used by some banks, are efficient and inexpensive, but they lack human judgment. They will dispense cash to anyone who inserts the right card and punches the right code into a keyboard.

Some of the more conventional methods of identification of personnel are covered in the discussion of access control in Chapter 10. These include special keys, encoded cards, and electronic combination locks. All of these measures are good and are probably adequate where the risk isn't too great. For example, if an automatic bank teller was set to dispense not more than $25.00 to any one depositor over a 24-hour period, the loss resulting from someone fooling the machine would not be severe. In highly sensitive situations, however, all of these means involve a certain amount

of risk. Keys and cards may be lost or stolen, and people are notoriously loose lipped about things like special code numbers.

Several electronic systems have been developed that will automatically identify a person on the basis of some physical characteristic that is not easily counterfeited.

ELECTRONIC FINGERPRINT IDENTIFICATION

It has long been recognized that the human fingerprint is a unique property of an individual. Fingerprints are used at all levels of law enforcement to recognize persons with criminal records. The reason that fingerprints have not been widely used to identify persons at security checkpoints or bank windows is that reading a fingerprint is not a simple task and requires special training.

One way of using a fingerprint for identification is to compare a person's fingerprint with a photograph of the fingerprint of an authorized person. With regular photography, this is impractical, but with a comparatively new technique—holography—it can and is being done.

Technically, holography is a technique for recording both the amplitude and the phase of an image, and later reconstructing the image. Fig. 14-1 is a sketch of the method of making a holographic plate. The light source is a laser which produces coherent light. The light reaches the photographic plate through two paths. Through one path, it is merely reflected from the reference mirror onto the plate. In the other path, light is reflected from the object being photographed to the photographic plate. The result is that an interference pattern is recorded on the plate. The plate itself can respond only to the amplitude or amount of light and not its phase. The interference pattern, however, contains the phase information as well. It usually bears little resemblance to the object

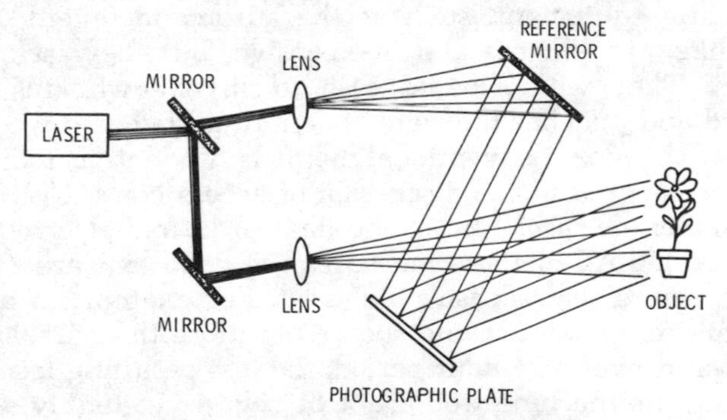

Fig. 14-1. The making of a hologram.

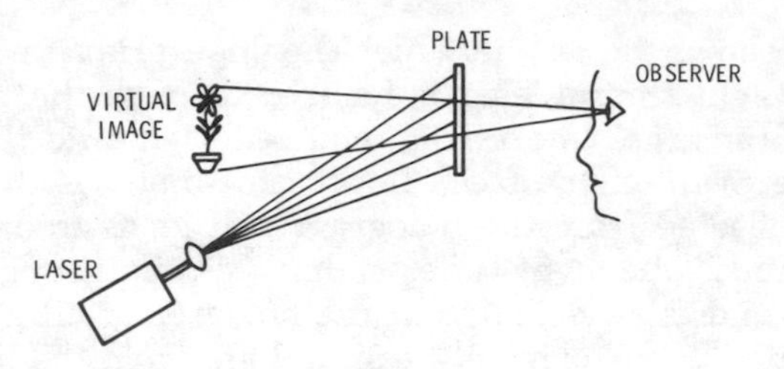

Fig. 14-2. Viewing a hologram.

being photographed but appears to be an irregular pattern of fringes.

When the transparency is illuminated by a laser, the image of the photographed object will be seen as shown in Fig. 14-2. There will be a virtual image that looks exactly like a three-dimensional image of the object. In fact, by moving slightly, the observer can actually see behind the object.

Fig. 14-3 shows an automatic personnel verifier based on the holographic principle. The person desiring to enter a facility carries a plastic card that has a holographic image of his or her fingerprint. He inserts the card into a slot and then inserts his finger into an opening as shown in the figure. A fresh image of his fingerprint is made on a glass plate, and this is compared with the hologram on the card. If the two prints match, the person is allowed access. If not, or if the device is tampered with, an alarm is initiated.

Fig. 14-3. Automatic personnel verifier.

Courtesy KMS Industries, Inc.

Fig. 14-4 shows the way in which the fingerprint is matched with its image. Light from a laser is focused through the fresh fingerprint onto a mirror, where it is then reflected onto the hologram of the fingerprint. If the fresh fingerprint matches the hologram, the light reflected from the hologram will focus to a bright spot. This is because the light, after it has passed through the fresh fingerprint and is reflected from the hologram, will actually be a reconstruction of the original beam of light from the laser used to make the hologram. If the fingerprint and hologram do not match, the light will be diffused. A photoelectric sensor is used to distinguish between the bright spot and the diffused light, and to trip an alarm if the fingerprint does not match the hologram on the card.

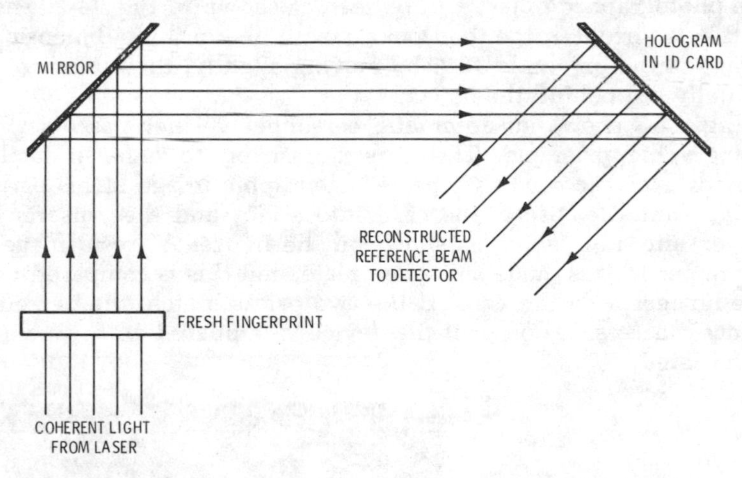

Fig. 14-4. Operating principles of automatic personnel verifier.

HAND GEOMETRY

Although it has been known for a long time that fingerprints are unique, it is also true that many other human characteristics are unique to the individual. An Air Force study of fitting high-altitude gloves to pilots nearly twenty-five years ago indicated that certain properties of the human hand such as the lengths of the fingers and palms were almost as unique as fingerprints.

This property has recently been used in a personnel identification system. In this system, certain parameters of the hands of authorized persons are encoded on a card. The card is placed in a slot, and the hand is placed on a lighted surface where the corresponding parameters are measured electronically. The two sets of

measurements are then compared in a digital circuit. If the card belongs to the person whose hand is on the screen, access is permitted. Otherwise an alarm is sounded.

To allow for errors, the device is usually set to allow a person to make a second try in case he didn't place his hand on the screen correctly. This reduces the number of false alarms but ensures that an unauthorized person is not allowed access.

VOICE PRINTS

Everyone is familiar with the fact that no two people have exactly the same voice. To someone trained in voice recognition, the differences are even greater. In fact, the testimony of persons trained in voice recognition has been accepted as positive identification against persons accused of illegally operating a radio transmitter.

Although a human listener can recognize hundreds of different voices with little effort, the problem of recognizing voices automatically is not a simple one. It has been done successfully in the laboratory but has not (as of this writing) been applied to identification equipment in the field.

One application of voice printing is the identification of anonymous telephone callers; a recording of the anonymous call is compared with recordings of the same statements by various persons suspected of making the call. At least one company is now offering this service to law enforcement agencies and security personnel.

With the advent of low-cost microprocessor systems, the use of voice-printing will undoubtedly grow in the near future.

CHAPTER 15

Verification

One of the oldest security problems is determining the truthfulness of testimony. Ancient methods for telling whether or not a witness was lying were far from adequate. In some of these testing procedures, both the innocent and the guilty perished from the procedure, thus eliminating the necessity of further investigation.

Much security and law enforcement work is based on information obtained from the testimony of various people. The selection of personnel for positions involving a high security risk is based to an amazing extent on statements made by them or on references provided by them. Much criminal investigation consists of following up clues that are contained in statements.

It is unfortunate that many human beings habitually treat the truth in a way that serves their personal interests. The exaggeration or the slanted statement is often more troublesome because it is only partially true, making an objective evaluation of the situation very difficult.

Many honest applicants have been denied employment because of a prejudicial character reference. An employee reporting on the circumstances of an intrusion may twist the facts not because he is guilty, but because honest testimony may direct attention to some other misdemeanor, such as neglect of his job. All of these compromises with truth only complicate the job of the security or law enforcement official.

Unfortunately, there is no electronic principle that can be utilized to directly tell the difference between true and false testimony. There are, however, many physiological parameters that

can be measured electronically to help a skilled operator distinguish between true and false statements.

The oldest device used to verify the truth of testimony is the polygraph, or lie detector. This device measures many physiological parameters such as heart rate and skin resistance, and plots them on a strip chart. Variations in these parameters are then correlated with answers to questions. For example, the palms of some people will perspire when they feel tension, i.e., when they might be lying. This slight increase in perspiration will be accompanied by a change in the resistance measured at the skin.

The polygraph has many limitations. First of all, it has to be connected to the individual. This in itself produces tension and complicates the job of interpreting the measurement. Second, it is often considered an invasion of privacy. The use of the polygraph usually requires the consent of the person being questioned.

THE PSYCHOLOGICAL STRESS EVALUATOR

A comparatively recent development that has many advantages over the polygraph is the *P*sychological *S*tress *E*valuator, or PSE. This device evaluates the psychological stress that a person experiences. It does this by making certain measurements from a tape recording of the person's voice. The PSE has the advantage of requiring no physical connection of the victim to the device. The device can be used without the knowledge of the victim, although it may not be legal to do so.

The operation of the psychological stress evaluator is based on the fact that there is normally a slight tremor—called a microtremor—in the muscles of the body, including the muscles that control the voice. Under psychological stress, this tremor is suppressed or completely eliminated. When the stress abates, the tremor returns.

The effect that the tremor has on the voice cannot be distinguished by the ear. When listening to a person speak, the listener cannot perceive the presence (or absence) of the tremor. The effect of the tremor is a small frequency modulation of about 10 Hz superimposed on the normal voice sounds.

This fm component on the voice can be measured by playing a tape recording of the voice through an analyzer similar to an fm discriminator and plotting the result on a strip chart as shown in Fig. 15-1. For more detailed analysis, the tape speed of the playback can be changed.

Fig. 15-2 shows two plots of the output of the analyzer. In Fig. 15-2A, there is a constant, nearly uniform, motion superimposed on the trace. In Fig. 15-2B, this tremor is missing at certain points

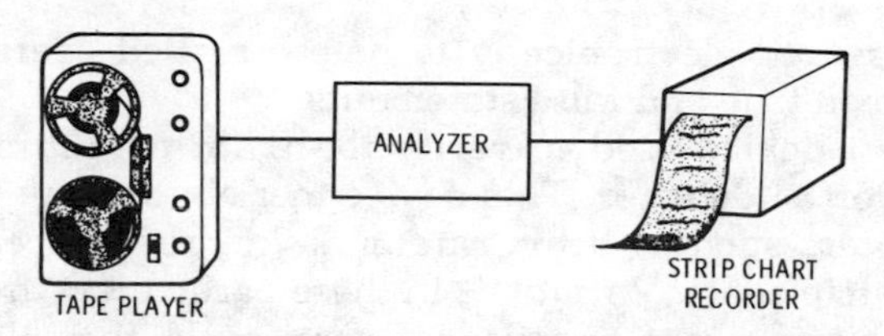

Fig. 15-1. Psychological stress evaluator (PSE).

on the trace. A skilled observer can readily detect the straight or diagonal lines that indicate psychological stress.

In general, the PSE requires the services of a trained operator for effective use. The phrasing of the questions asked by the questioner is very important. It is possible for an unskilled questioner to induce stress in an innocent party. For example, if when questioning a suspect about an act, the questioner brings to mind another act for which the suspect was guilty in the past, stress may result.

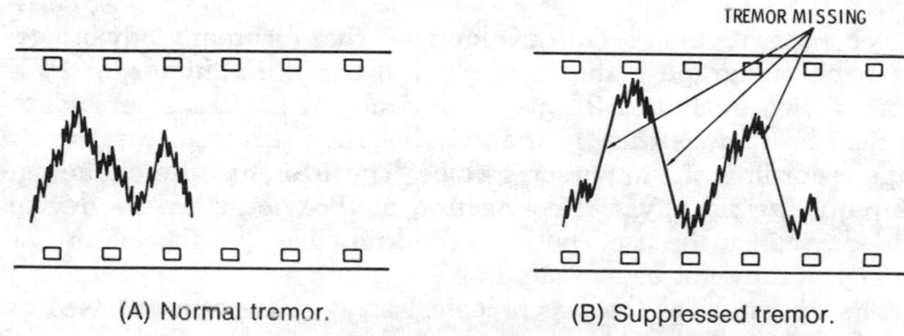

(A) Normal tremor. (B) Suppressed tremor.

Fig. 15-2. Plots from psychological stress evaluator.

The PSE is already used by many law enforcement agencies and security personnel. It has value not only in investigating a crime, but also in questioning applicants for sensitive positions before they are hired. Unfortunately, it also has possibilities for use in industrial espionage. The fact that tape recordings can be made of telephone conversations means that the technique can be used on anyone without his knowledge. False statements made to protect business statements may be detected, and slowly the secret that is supposed to be concealed is revealed.

CHAPTER 16

Computers in Security Systems

In Chapter 12 the increasing number of crimes committed in connection with a computer were discussed. These crimes included the destruction of data held in the computer as well as the diversion of funds or data controlled by the computer. The computer, however, is not just a vehicle for crime; it can also constitute an effective security system against crime.

The computer has three principal characteristics that make it well suited to use in security systems:

1. A computer can handle vast amounts of data in a very short period of time.
2. Computers can be programmed to make decisions. Unlike a conventional system that sounds an alarm when a certain signal level is reached, a computer can evaluate alarm signals to give a better indication of the situation.

Computer-based systems will not replace the detectors described in earlier chapters. Instead, they process the data received from ordinary intrusion detectors. Because of the tremendous data-handling capability of a computer, it can perform many other functions simultaneously with security monitoring. Thus, the cost of the system apportioned to security functions is reduced.

THE MICROPROCESSOR

Of course, computers have been available for many years and have been used for controlling security systems in some very large

facilities. However, until recently their high cost has kept them from being used in any but the most elaborate computer security systems.

The recent introduction of the microprocessor and associated integrated circuits has changed this situation drastically. It is now possible to build a computer-based security system for a fraction of the former cost.

The microprocessor is simply an integrated circuit that contains most or all of the control processing unit of a digital computer. The cost-reduction possibilities incidental to integrated-circuit technology have brought the cost of a microprocessor down to a few dollars. Complete microcomputer systems are now priced under $1000.00.

The principal problems encountered in implementing computer-based security systems are not found in computer technology but in the peripheral devices (such as printers) used to get the data into and out of the system.

The microprocessor has another advantage in security systems, the advantage of adaptability. That is, most of the functions that a microprocessor performs are controlled by binary signals stored in its memory, and these signals are easily changed. The program of a computer is much like the block diagram of a conventional piece of equipment. In a conventional system, it is necessary to rewire the system to change the function of the system or any part of it. With a computer-based system, it is necessary only to change the binary signals stored in its memory.

Security systems tend to be somewhat unique. No two complete systems are exactly alike. This means that as far as the installation of the system is concerned, some design work is required to develop the final wiring diagram. A single computer-based system, on the other hand, can be adapted to many different applications without changing any of the wiring; it is necessary only to change the program stored in the memory.

A CITY-WIDE PROTECTION SYSTEM

Fig. 16-1 shows a block diagram of a recently introduced system for security protection. The area to be protected, which might be anything from an industrial complex or an apartment building to a single home, is equipped with intrusion detectors. Each terminal contains a semiconductor memory that identifies the building or home by its full address and even identifies each entrance. When an intrusion occurs, the device will automatically dial a predetermined number at police headquarters or at a security company and will then initiate a printout giving the address

of the facility where the alarm originated and the nature of the alarm.

Because of the flexibility of the digital techniques, some data processing can be performed before the alarm is initiated. For example, if the alarm system guarding the outer perimeter of the premises is tripped, the system will transmit a signal that lets the police know exactly what has happened—the outer perimeter has been penetrated. Other alarms inside the perimeter, such as pressure pads, will be tripped when the burglar reaches this point and another more-urgent alarm will be transmitted.

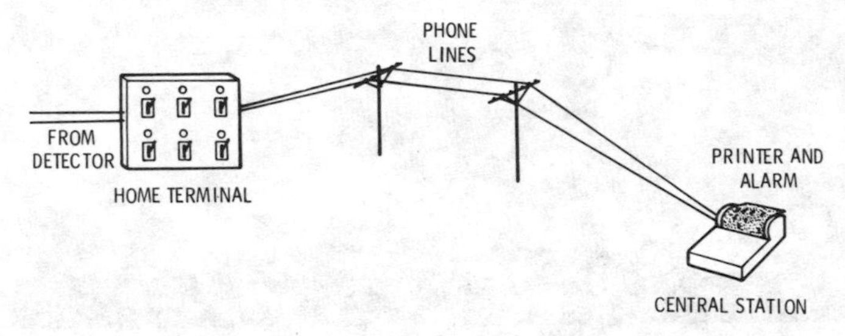

Fig. 16-1. Pan-X city-wide protection system.

With proper design and installation, the probability of false alarms will be minimized. When the receiving facility at police headquarters receives several alarms that trace the progress of an intruder through the premises, the police can be almost 100% certain that an intrusion is taking place and that it is not a false alarm.

Using modern integrated circuits, the amount of equipment needed to implement this system is very little, and the cost is well within the reach of the average homeowner. Fig. 16-2 shows the printed-circuit board containing the system.

Because of the tremendous data-handling capability of the system, it can also perform many functions in the home other than security monitoring. The details of these are beyond the scope of this book. One function that is related to security, however, is the holdup alarm.

When a home is occupied during the daytime, a security system will usually be turned off to minimize false alarms. Under these circumstances, an intruder could force his way into the house without tripping an alarm. To combat this possibility, the system is equipped with a holdup alarm. Here a small radio transmitter, no larger than a pack of cigarettes, carries a panic button. When the button is pressed, a signal is sent to a local receiver which then directs the system to transmit a holdup alarm to police headquar-

Courtesy Pan-X Corp.

Fig. 16-2. Printed-circuit board used in home terminal.

ters. Again the message contains the address as well as a code indicating that the emergency is a holdup.

APARTMENT AND OFFICE COMPLEXES

Large apartment and office complexes have been traditionally protected by guards patrolling the facilities. This system has limitations in that guards cannot be at all parts of the facility at the same time. By using computer-based systems, a tremendous number of intrusion detectors can be monitored and the information relayed to a central facility from which guards can be dispatched to where they are needed.

Many hotels and office buildings are now using computers in their security systems. A typical system may have as many as 1000 intrusion detectors of various types connected to the computer. One of the advantages of the system is that detectors of different types can be used in the same system. When cash drawers are unattended, they can be protected by simple switches that will alert the system if the drawer is opened. Holdup alarms can be provided for cash positions that are in use. Motion detectors can be installed on the roof and in areas that are not normally occupied.

Because of the decision-making capacity of the computer, many functions can be performed automatically. For example, the system can arm the protective circuits in offices 30 minutes after normal office hours are over. When this occurs, the location of any unlocked doors or occupied offices will be automatically typed out on the printer at guard headquarters. If a person intends to work late, the guard headquarters can be notified in advance. Thus, a detailed record is provided of all activities in the facility after normal business hours. Periodic inspection of this record will improve overall security.

The rapid decrease in the costs of microprocessors and associated equipment will soon bring computer-based systems into almost every home. Systems of this type will be exceptionally hard to foil and will constitute a significant deterrent to burglary.

"SMART" DETECTORS

The low cost of the microprocessor is leading to the development of what are often called "smart" intrusion detectors. The smart detector is based on the ability of a computer to look at a large amount of data, compare it with data in a memory, and decide whether the two groups of data agree.

The principle of a smart detector can be illustrated by the electronic line-scan camera shown in Fig. 16-3. This camera is similar to an ordinary camera except that the space where the film would

Courtesy Recticon Corp.

Fig. 16-3. Electronic line-scan camera.

be in an ordinary camera is occupied by an array of tiny photodiodes. The photodiodes are scanned linearly, producing an analog signal that is a function of the amount of light at the various points in the array. This signal is compared with preset signal levels to form a digital signal corresponding to the difference in the actual light level at each point on the scene and the preset level. A logical zero is produced for light levels that are below the preset level and a logical one for light levels that exceed the preset level.

By programming the light levels corresponding to a given scene, the device cannot only detect a change in the scene, but can also tell something about the nature of a change. For example, the signal could tell if an object moved to the right or to the left and by how much. Since the output is a digital signal, memories can be programmed to print out the nature of the change.

This general concept of using the data-handling capabilities of a microprocessor need not be limited to visual images, although this is a fruitful field of application. It is possible to construct multimode sensors that will establish the nature of an intruder. Infrared sensors can be combined with pressure pads to compute whether or not the intruder is a human being and how many intruders are in an area.

The field of "smart" detectors is in its infancy. It can be expected that the entire field of security electronics will be significantly affected by it in the near future.

APPENDIX **A**

Glossary

This list of terms and definitions applicable to security systems was prepared by the Law Enforcement Standards Laboratory which has been established at the United States National Bureau of Standards.

Access Control—The control of pedestrian and vehicular traffic through entrances and exits of a protected area or premises.

Access Mode—The operation of an alarm system such that no alarm signal is given when the protected area is entered; however, a signal may be given if the sensor, annunciator, or control unit is tampered with or opened.

Access/Secure Control Unit—*See* Control Unit.

Access Switch—*See* Authorized Access Switch.

Accumulator—A circuit which accumulates a sum. For example, in an audio alarm control unit, the accumulator sums the amplitudes of a series of pulses, which are larger than some threshold level, subtracts from the sum at a predetermined rate to account for random background pulses, and initiates an alarm signal when the sum exceeds some predetermined level. This circuit is also called an integrator; in digital circuits it may be called a counter.

Active Intrusion Sensor—An active sensor which detects the presence of an intruder within the range of the sensor. Examples are an ultrasonic motion detector, a radio frequency motion detector, and a photoelectric alarm system. *See also* Passive Intrusion Sensor.

Active Sensor—A sensor which detects the disturbance of a radiation field which is generated by the sensor. *See also* Passive Sensor.

Actuating Device—*See* Actuator.

Actuator—A manual or automatic switch or sensor such as a holdup button, magnetic switch, or thermostat which causes a system to transmit an alarm signal when manually activated or when the device automatically senses an intruder or other unwanted condition.

Air Gap—The distance between two magnetic elements in a magnetic or electromagnetic circuit, such as between the core and the armature of a relay.

Alarm—An alarm device or an alarm signal.

Alarm Condition—A threatening condition, such as an intrusion, fire, or holdup, sensed by a detector.

Alarm Device—A device which signals a warning in response to an alarm condition, such as a bell, siren, or annunciator.

Alarm Discrimination—The ability of an alarm system to distinguish between those stimuli caused by an intrusion and those which are a part of the environment.

Alarm Line—A wired electrical circuit used for the transmission of alarm signals from the protected premises to a monitoring station.

Alarm Receiver—*See* Annunciator.

Alarm Sensor—*See* Sensor.

Alarm Signal—A signal produced by a control unit indicating the existence of an alarm condition.

Alarm State—The condition of a detector which causes a control unit in the secure mode to transmit an alarm signal.

Alarm Station—(1) A manually actuated device installed at a fixed location to transmit an alarm signal in response to an alarm condition, such as a concealed holdup button in a bank teller's cage. (2) A well-marked emergency control unit, installed in fixed locations usually accessible to the public, used to summon help in response to an alarm condition. The control unit contains either a manually actuated switch or telephone connected to fire or police headquarters, or a telephone answering service. *See also* Remote Station Alarm System.

Alarm System—An assembly of equipment and devices designated and arranged to signal the presence of an alarm condition requiring urgent attention such as unauthorized entry, fire, temperature rise, etc. The system may be local, police connection, central station or proprietary. (For individual alarm systems see alphabetical listing by type, e.g., Intrusion Alarm System.)

Annunciator—An alarm monitoring device which consists of a number of visible signals such as "flags" or lamps indicating the status of the detectors in an alarm system or systems. Each circuit in the device is usually labeled to identify the location and condition being monitored. In addition to the visible signal, an audible signal is usually associated with the device. When an alarm condition is reported, a signal is indicated visibly, audibly, or both. The visible signal is generally maintained until reset either manually or automatically.

Answering Service—A business which contracts with subscribers to answer incoming telephone calls after a specified delay or when scheduled to do so. It may also provide other services such as relaying fire or intrusion alarm signals to proper authorities.

Area Protection—Protection of the inner space or volume of a secured area by means of a volumetric sensor.

Area Sensor—A sensor with a detection zone which approximates an area, such as a wall surface or the exterior of a safe.

Audible Alarm Device—(1) A noisemaking device such as a siren, bell, or horn used as part of a local alarm system to indicate an alarm condition. (2) A bell, buzzer, horn or other noisemaking device used as a part of an annunciator to indicate a change in the status or operating mode of an alarm system.

Audio Detection System—*See* Sound Sensing Detection System.

Audio Frequency (Sonic)—Sound frequencies within the range of human hearing, approximately 15 to 20,000 Hz.

Audio Monitor—An arrangement of amplifiers and speakers designed to monitor the sounds transmitted by microphones located in the protected

area. Similar to an annunciator, except that supervisory personnel can monitor the protected area to interpret the sounds.

Authorized Access Switch—A device used to make an alarm system or some portion or zone of a system inoperative in order to permit authorized access through a protected port. A shunt is an example of such a device.

B. A.—Abbreviation for burglar alarm.

Beam Divergence—In a photoelectric alarm system, the angular spread of the light beam.

Break Alarm—(1) An alarm condition signaled by the opening or breaking of an electrical circuit. (2) The signal produced by a break alarm condition (sometimes referred to as an open circuit alarm or trouble signal, designed to indicate possible system failure).

Bug—(1) To plant a microphone or other sound sensor or to tap a communication line for the purpose of surreptitious listening or audio monitoring; loosely, to install a sensor in a specified location. (2) The microphone or other sensor used for the purpose of surreptitious listening.

Building Security Alarm System—The system of protective signaling devices installed at a premise.

Burglar Alarm (*B. A.*) *Pad*—A supporting frame laced with fine wire or a fragile panel located with foil or fine wire and installed so as to cover an exterior opening in a building, such as a door, or skylight. Entrance through the opening breaks the wire or foil and initiates an alarm signal. *See also* Grid.

Burglar Alarm System.—*See* Intrusion Alarm System.

Burglary—The unlawful entering of a structure with the intent to commit a felony or theft therein.

Cabinet-for-Safe—A wooden enclosure having closely spaced electrical grids on all inner surfaces and contacts on the doors. It surrounds a safe and initiates an alarm signal if an attempt is made to open or penetrate the cabinet.

Capacitance—The property of two or more objects which enables them to store electrical energy in an electric field between them. The basic measurement unit is the farad. Capacitance varies inversely with the distance between the objects; hence the change of capacitance with relative motion is greater the nearer one object is to the other.

Capacitance Alarm System—An alarm system in which a protected object is electrically connected as a capacitance sensor. The approach of an intruder causes sufficient change in capacitance to upset the balance of the system and initiate an alarm signal. Also called proximity alarm system.

Capacitance Detector—*See* Capacitance Sensor.

Capacitance Sensor—A sensor which responds to a change in capacitance in a field containing a protected object or in a field within a protected area.

Carrier Current Transmitter—A device which transmits alarm signals from a sensor to a control unit via the standard ac power lines.

Central Station—A control center to which alarm systems in a subscriber's premises are connected, where circuits are supervised, and where personnel are maintained continuously to record and investigate alarm or trouble signals. Facilities are provided for the reporting of alarms to police and fire departments or to other outside agencies.

Central Station Alarm System—An alarm system, or group of systems, the activities of which are transmitted to, recorded in, maintained by, and supervised from a central station. This differs from proprietary alarm systems in that the central station is owned and operated independently of the subscriber.

Circumvention—The defeat of an alarm system by the avoidance of its de-

tection devices, such as by jumping over a pressure sensitive mat, by entering through a hole cut in an unprotected wall rather than through a protected door, or by keeping outside the range of an ultrasonic motion detector. Circumvention contrasts with spoofing.

Closed Circuit Alarm—See Cross Alarm.

Closed Circuit System—A system in which the sensors of each zone are connected in series so that the same current exists in each sensor. When an activated sensor breaks the circuit or the connecting wire is cut, and alarm is transmitted for that zone.

Clutch Head Screw—A mounting screw with a uniquely designed head for which the installation and removal tool is not commonly available. They are used to install alarm system components so that removal is inhibited.

Coded-Alarm System—An alarm system in which the source of each signal is identifiable. This is usually accomplished by means of a series of current pulses which operate audible or visible annunciators or recorders, or both, to yield a recognizable signal. This is usually used to allow the transmission of multiple signals on a common circuit.

Coded Cable—A multiconductor cable in which the insulation on each conductor is distinguishable from all others by color or design. This assists in identification of the point of origin or final destination of a wire.

Coded Transmitter—A device for transmitting a coded signal when manually or automatically operated by an actuator. The actuator may be housed with the transmitter or a number of actuators may operate a common transmitter.

Coding Siren—A siren which has an auxiliary mechanism to interrupt the flow of air through its principal mechanism, enabling it to produce a controllable series of sharp blasts.

Combination Sensor Alarm System—An alarm system which requires the simultaneous activation of two or more sensors to initiate an alarm signal.

Compromise—See Defeat.

Constant Ringing Drop (*CRD*)—A relay which, when activated even momentarily, will remain in an alarm condition until reset. A key is often required to reset the relay and turn off the alarm.

Constant Ringing Relay (*CRR*)—*See* Constant Ringing Drop.

Contact—(1) Each of the pair of metallic parts of a switch or relay which by touching or separating make or break the electrical current path. (2) A switch-type sensor.

Contact Device—A device which when actuated opens or closes a set of electrical contacts; a switch or relay.

Contact Microphone—A microphone designed for attachment directly to a surface of a protected area or object; usually used to detect surface vibrations.

Contact Vibration Sensor—See Vibration Sensor.

Contactless Vibrating Bell—A vibrating bell whose continuous operation depends upon application of an alternating current, without circuit-interrupting contacts such as those used in vibrating bells operated by direct current.

Control Cabinet—See Control Unit.

Control Unit—A device, usually electronic, which provides the interface between the alarm system and the human operator and produces an alarm signal when its programmed response indicates an alarm condition. Some or all of the following may be provided for: power for sensors, sensitivity adjustments, means to select and indicate access mode or secure mode, monitoring for line supervision and tamper devices, timing circuit for entrance and exit delays, transmission of an alarm signal, etc.

Covert—Hidden and protected.

CRD—*See* Constant Ringing Drop.

Cross Alarm—(1) An alarm condition signaled by crossing or shorting an electrical circuit. (2) The signal produced due to a cross alarm condition.

Crossover—An insulated electrical path used to connect foil across window dividers, such as those found on multiple pane windows, to prevent grounding and to make a more durable connection.

CRR—Constant ringing relay. *See* Constant Ringing Drop.

Dark Current—The current output of a photoelectric sensor when no light is entering the sensor.

Day Setting—*See* Access Mode.

Defeat—The frustration, counteraction, or thwarting of an alarm device so that it fails to signal an alarm when a protected area is entered. Defeat includes both circumvention and spoofing.

Detection Range—The greatest distance at which a sensor will consistently detect an intruder under a standard set of conditions.

Detector—(1) A sensor such as those used to detect intrusion, equipment malfunctions or failure, rate of temperature rise, smoke or fire. (2) A demodulator, a device for recovering the modulating function or signal from a modulated wave, such as that used in a modulated photoelectric alarm system. *See also* Photoelectric Alarm System, Modulated.

Dialer—*See* Telephone Dialer, Automatic.

Differential Pressure Sensor—A sensor used for perimeter protection which responds to the difference between the hydraulic pressures in two liquid-filled tubes buried just below the surface of the earth around the exterior perimeter of the protected area. The pressure difference can indicate an intruder walking or driving over the buried tubes.

Digital Telephone Dialer—*See* Telephone Dialer, Digital.

Direct Connect—*See* Police Connection.

Direct Wire Burglar Alarm Circuit (DWBA)—*See* Alarm Line.

Direct Wire Circuit—*See* Alarm Line.

Door Cord—A short, insulated cable with an attaching block and terminals at each end used to conduct current to a device, such as foil, mounted on the movable portion of a door or window.

Door Trip Switch—A mechanical switch mounted so that movement of the door will operate the switch.

Doppler Effect (Shift)—The apparent change in frequency of sound or radio waves when reflected from or originating from a moving object. Utilized in some types of motion sensors.

Double-Circuit System—An alarm circuit in which two wires enter and two wires leave each sensor.

Double Drop—An alarm signaling method often used in central station alarm systems in which the line is first opened to produce a break alarm and then shorted to produce a cross alarm.

Drop—(1) *See* Annunciator. (2) A light indicator on an annunciator.

Duress Alarm Device—A device which produces either a silent alarm or local alarm under a condition of personnel stress such as holdup, fire, illness, or other panic or emergency. The device is normally manually operated and may be fixed or portable.

Duress Alarm System—An alarm system which employs a duress alarm device.

DWBA—Direct wire burglar alarm. *See* Alarm Line.

E-Field Sensor—A passive sensor which detects changes in the earth's ambient electric field caused by the movement of an intruder. *See also* H-Field Sensor.

Electrical—Related to, pertaining to, or associated with electricity.

Electromagnetic—Pertaining to the relationship between current flow and magnetic field.

Electromagnetic Interference (EMI)—Impairment of the reception of a wanted electromagnetic signal by an electromagnetic disturbance. This can be caused by lightning, radio transmitters, power line noise and other electrical devices.

Electromechanical Bell—A bell with a prewound spring-driven striking mechanism, the operation of which is initiated by the activation of an electric tripping mechanism.

Electronic—Related to, or pertaining to, devices which utilize electrons moving through a vacuum, gas, or semiconductor, and to circuits or systems containing such devices.

EMI—See Electromagnetic Interference.

End of Line Resistor—See Terminal Resistor.

Entrance Delay—The time between actuating a sensor on an entrance door or gate and the sounding of a local alarm or transmission of an alarm signal by the control unit. This delay is used if the authorized access switch is located within the protected area and permits a person with the control key to enter without causing an alarm. The delay is provided by a timer within the control unit.

E.O.L.—End of line.

Exit Delay—The time between turning on a control unit and the sounding of a local alarm or transmission of an alarm signal upon actuation of a sensor on an exit door. This delay is used if the authorized access switch is located within the protected area and permits a person with the control key to turn on the alarm system and to leave through a protected door or gate without causing an alarm. The delay is provided by a timer within the control unit.

Fail Safe—A feature of a system or device which initiates an alarm or trouble signal when the system or device either malfunctions or loses power.

False Alarm—An alarm signal transmitted in the absence of an alarm condition. These may be classified according to causes: environmental, e.g., rain, fog, wind, hail, lightning, temperature, etc.; animals, e.g., rats, dogs, cats, insects, etc.; man-made disturbances, e.g., sonic booms, EMI, vehicles, etc.; equipment malfunction, e.g., transmission errors, component failure, etc.; operator error; and unknown.

False Alarm Rate, Monthly—The number of false alarms per installation per month.

False Alarm Ratio—The ratio of false alarms to total alarms; may be expressed as a percentage or as a simple ratio.

Fence Alarm—Any of several types of sensors used to detect the presence of an intruder near a fence or any attempt by him to climb over, go under, or cut through the fence.

Field—The space or area in which there exists a force such as that produced by an electrically charged object, a current, or a magnet.

Fire Detector (Sensor)—See Heat Sensor and Smoke Detector.

Floor Mat—See Mat Switch.

Floor Trap—A trap installed so as to detect the movement of a person across a floor space, such as a trip wire switch or mat switch.

Foil—Thin metallic strips which are cemented to a protected surface (usually glass in a window or door), and connected to a closed electrical circuit. If the protected material is broken so as to break the foil, the circuit opens, initiating an alarm signal. Also called tape. A window, door, or other surface to which foil has been applied is said to be taped or foiled.

Foil Connector—An electrical terminal block used on the edge of a window to join interconnecting wire to window foil.

Foot Rail—A holdup alarm device, often used at cashiers' windows, in which a foot is placed under the rail, lifting it, to initiate an alarm signal.

Frequency Division Multiplexing (FDM)—See Multiplexing, Frequency Division.

Glassbreak Vibration Detector—A vibration detection system which employs a contact microphone attached to a glass window to detect cutting or breakage of the glass.

Grid—(1) An arrangement of electrically conducting wire, screen, or tubing placed in front of doors or windows or both which is used as a part of a capacitance sensor. (2) A lattice of wooden dowels or slats concealing fine wires in a closed circuit which initiates an alarm signal when forcing or cutting the lattice breaks the wires. Used over accessible openings. Sometimes called a protective screen. *See also* Burglar Alarm Pad. (3) A screen or metal plate, connected to earth ground, sometimes used to provide a stable ground reference for objects protected by a capacitance sensor. If placed against the walls near the protected object, it prevents the sensor sensitivity from extending through the walls into areas of activity.

Heat Detector—See Heat Sensor.

Heat Sensor—(1) A sensor which responds to either a local temperature above a selected value, a local temperature increase which is at a rate of increase greater than a preselected rate (rate of rise), or both. (2) A sensor which responds to infrared radiation from a remote source such as a person.

H-Field Sensor—A passive sensor which detects changes in the earth's ambient magnetic field caused by the movement of an intruder. *See also* E-Field Sensor.

Holdup—A robbery involving the threat to use a weapon.

Holdup Alarm Device—A device which signals a holdup. The device is usually surreptitious and may be manually or automatically actuated, fixed or portable. *See* Duress Alarm Device.

Holdup Alarm System, Automatic—An alarm system which employs a holdup alarm device, in which the signal transmission is initiated solely by the action of the intruder, such as a money clip in a cash drawer.

Holdup Alarm System, Manual—A holdup alarm system in which the signal transmission is initiated by the direct action of the person attacked or of an observer of the attack.

Holdup Button—A manually actuated mechanical switch used to initiate a duress alarm signal; usually constructed to minimize accidental activation.

Hood Contact—A switch which is used for the supervision of a closed safe or vault door. Usually installed on the outside surface of the protected door.

Impedance—The opposition to the flow of alternating current in a circuit. May be determined by the ratio of an input voltage to the resultant current.

Impedance Matching—Making the impedance of a terminating device equal to the impedance of the circuit to which it is connected in order to achieve optimum signal transfer.

Infrared (IR) Motion Detector—A sensor which detects changes in the infrared light radiation from parts of the protected area. Presence of an intruder in the area changes the infrared light intensity from his direction.

Infrared (IR) Motion Sensor—See Infrared Motion Detector.

Infrared Sensor—See Heat Sensor, Infrared Motion Detector, and Photoelectric Sensor.

Inking Register—See Register, Inking.

Interior Perimeter Protection—A line of protection along the interior bound-

ary of a protected area including all points through which entry can be effected.

Intrusion—Unauthorized entry into the property of another.

Intrusion Alarm System—An alarm system for signaling the entry or attempted entry of a person or an object into the area or volume protected by the system.

Ionization Smoke Detector—A smoke detector in which a small amount of radioactive material ionizes the air in the sensing chamber, thus rendering it conductive and permitting a current to flow through the air between two charged electrodes. This effectively gives the sensing chamber an electrical conductance. When smoke particles enter the ionization area, they decrease the conductance of the air by attaching themselves to the ions causing a reduction in mobility. When the conductance is less than a predetermined level, the detector circuit responds.

IR—Infrared.

Jack—An electrical connector which is used for frequent connect and disconnect operations; for example, to connect an alarm circuit at an overhang door.

Lacing—A network of fine wire surrounding or covering an area to be protected, such as a safe, vault, or glass panel, and connected into a closed circuit system. The network of wire is concealed by a shield such as concrete or paneling in such a manner that an attempt to break through the shield breaks the wire and initiates an alarm.

Light Intensity Cutoff—In a photoelectric alarm system, the percent reduction of light which initiates an alarm signal at the photoelectric receiver unit.

Line Amplifier—An audio amplifier which is used to provide preamplification of an audio alarm signal before transmission of the signal over an alarm line. Use of an amplifier extends the range of signal transmission.

Line Sensor (Detector)—A sensor with a detection zone which approximates a line or series of lines, such as a photoelectric sensor which senses a direct or reflected light beam.

Line Supervision—Electronic protection of an alarm line accomplished by sending a continuous or coded signal through the circuit. A change in the circuit characteristics, such as a change in impedance due to the circuit's having been tampered with, will be detected by a monitor. The monitor initiates an alarm if the change exceeds a predetermined amount.

Local Alarm—An alarm which when activated makes a loud noise (*see* Audible Alarm Device) at or near the protected area or floods the site with light, or both.

Local Alarm System—An alarm system which when activated produces an audible or visible signal in the immediate vicinity of the protected premises or object. This term usually applies to systems designed to provide only a local warning of intrusion and not to transmit to a remote monitoring station. However, local alarm systems are sometimes used in conjunction with a remote alarm.

Loop—An electric circuit consisting of several elements, usually switches, connected in series.

Magnetic Alarm System—An alarm system which will initiate an alarm when it detects changes in the local magnetic field. The changes could be caused by motion of ferrous objects such as guns or tools near the magnetic sensor.

Magnetic Contact—*See* Magnetic Switch.

Magnetic Sensor—A sensor which responds to changes in the magnetic field. *See also* Magnetic Alarm System.

Magnetic Switch—A switch which consists of two separate units: a magnetically actuated switch, and a magnet. The switch is usually mounted in a fixed position (door jamb or window frame) opposing the magnet, which is fastened to a hinged or sliding door, window, etc. When the movable section is opened, the magnet moves with it, actuating the switch.

Magnetic Switch, Balanced—A magnetic switch which operates using a balanced magnetic field in such a manner as to resist defeat with an external magnet. It signals an alarm when it detects either an increase or decrease in magnetic field strength.

Matching Network—A circuit used to achieve impedance matching. It may also allow audio signals to be transmitted to an alarm line while blocking direct current used locally for line supervision.

Mat Switch—A flat area switch used on open floors or under carpeting. It may be sensitive over an area of a few square feet or several square yards.

McCulloh Circuit (Loop)—A supervised single wire loop connecting a number of coded transmitters located in different protected areas to a central station receiver.

Mechanical Switch—A switch in which the contacts are opened and closed by means of a depressable plunger or button.

Mercury Fence Alarm—A type of mercury switch which is sensitive to the vibration caused by an intruder climbing on a fence.

Mercury Switch—A switch operated by tilting or vibrating which causes an enclosed pool of mercury to move, making or breaking physical and electrical contact with conductors. They are used on tilting doors and windows, and on fences.

Microwave Alarm System—An alarm system which employs radio frequency motion detectors operating in the microwave frequency region of the electromagnetic spectrum.

Microwave Frequency—Radio frequencies in the range of approximately 1.0 to 300 GHz.

Microwave Motion Detector—*See* Radio Frequency Motion Detector.

Modulated Photoelectric Alarm System—*See* Photoelectric Alarm System, Modulated.

Monitor Cabinet—An enclosure which houses the annunciator and associated equipment.

Monitor Panel—*See* Annunciator.

Monitoring Station—The central station or other area at which guards, police, or commercial service personnel observe the annunciators and the registers reporting on the condition of alarm systems.

Motion Detection System—*See* Motion Sensor.

Motion Detector—*See* Motion Sensor.

Motion Sensor—A sensor which responds to the motion of an intruder. *See also* Radio Frequency Motion Detector, Sonic Motion Detector, Ultrasonic Motion Detector, and Infrared Motion Detector.

Multiplexing—A technique for the concurrent transmission of two or more signals in either or both directions, over the same wire, carrier, or other communication channel. The two basic multiplexing techniques are time division multiplexing and frequency division multiplexing.

Multiplexing, Frequency Division (FDM)—The multiplexing technique which assigns to each signal a specific set of frequencies (called a channel) within the larger block of frequencies available on the main transmission path in much the same way that many radio stations broadcast two programs at the same time that can be separately received.

Multiplexing, Time Division (TDM)—The multiplexing technique which provides for the independent transmission of several pieces of information on

a time-sharing basis by sampling, at frequent intervals, the data to be transmitted.

Neutralization—See Defeat.

Nicad—(Contraction of "nickel cadmium.") A high performance, long-lasting rechargeable battery, with electrodes made of nickel and cadmium. Such batteries may be used as an emergency power supply for an alarm system.

Night Setting—See Secure Mode.

Nonretractable (One-Way) Screw—A screw with a head designed to permit installation with an ordinary flat bit screwdriver but which resists removal. They are used to install alarm system components so that removal is inhibited.

Normally Closed (NC) Switch—A switch in which the contacts are closed when no external forces act upon the switch.

Normally Open (NO) Switch—A switch in which the contacts are open (separated) when no external forces act upon the switch.

Nuisance Alarm—See False Alarm.

Object Protection—See Spot Protection.

Open-Circuit Alarm—See Break Alarm.

Open-Circuit System—A system in which the sensors are connected in parallel. When a sensor is activated, the circuit is closed, permitting a current which activates an alarm signal.

Panic Alarm—See Duress Alarm Device.

Panic Button—See Duress Alarm Device.

Passive Intrusion Sensor—A passive sensor in an intrusion alarm system which detects an intruder within the range of the sensor. Examples are a sound sensing detection system, a vibration detection system, an infrared motion detector, and an E-field sensor.

Passive Sensor—A sensor which detects natural radiation or radiation disturbances, but does not itself emit the radiation on which its operation depends.

Passive Ultrasonic Alarm System—An alarm system which detects the sounds in the ultrasonic frequency range caused by an attempted forcible entry into a protected structure. The system consists of microphones, a control unit containing an amplifier, filters, an accumulator, and a power supply. The unit's sensitivity is adjustable so that ambient noises or normal sounds will not initiate an alarm signal; however, noise above the preset level or a sufficient accumulation of impulses will initiate an alarm.

Percentage Supervision—A method of line supervision in which the current in, or resistance of, a supervised line is monitored for changes. When the change exceeds a selected percentage of the normal operating current or resistance in the line, an alarm signal is produced.

Perimeter Alarm Systm—An alarm system which provides perimeter protection.

Perimeter Protection—Protection of access to the outer limits of a protected area, by means of physical barriers, sensors on physical barriers, or exterior sensors not associated with a physical barrier.

Permanent Circuit—An alarm circuit which is capable of transmitting an alarm signal whether the alarm control is in access mode or secure mode. Used, for example, on foiled fixed windows, tamper switches, and supervisory lines. *See also* Supervisory Alarm System, Supervisory Circuit, and Permanent Protection.

Permanent Protection—A system of alarm devices such as foil, burglar alarm pads, or lacings connected in a permanent circuit to provide protection whether the control unit is in the access mode or secure mode.

Photoelectric Alarm System—An alarm system which employs a light beam

and photoelectric sensor to provide a line of protection. Any interruption of the beam by an intruder is sensed by the sensor. Mirrors may be used to change the direction of the beam. The maximum beam length is limited by many factors, some of which are the light source intensity, number of mirror reflections, detector sensitivity, beam divergence, fog, and haze.

Photoelectric Alarm System, Modulated—A photoelectric alarm system in which the transmitted light beam is modulated in a predetermined manner and in which the receiving equipment will signal an alarm unless it receives the properly modulated light.

Photoelectric Beam-Type Smoke Detector—A smoke detector which has a light source which projects a light beam across the area to be protected onto a photoelectric cell. Smoke between the light source and the receiving cell reduces the light reaching the cell, causing actuation.

Photoelectric Detector—See Photoelectric Sensor.

Photoelectric Sensor—A device which detects a visible or invisible beam of light and responds to its complete or nearly complete interruption. *See also* Photoelectric Alarm System and Photoelectric Alarm System, Modulated.

Photoelectric Spot-Type Smoke Detector—A smoke detector which contains a chamber with covers that prevent the entrance of light but allow the entrance of smoke. The chamber contains a light source and a photosensitive cell so placed that light is blocked from it. When smoke enters, the smoke particles scatter and reflect the light into the photosensitive cell, causing an alarm.

Point Protection—See Spot Protection.

Police Connection—The direct link by which an alarm system is connected to an annunciator installed in a police station. Examples of a police connection are an alarm line and a radiocommunications channel.

Police Panel—See Police Station Unit.

Police Station Unit—An annunciator which can be placed in operation in a police station.

Portable Duress Sensor—A device carried on a person which may be activated in an emergency to send an alarm signal to a monitoring station.

Portable Intrusion Sensor—A sensor which can be installed quickly and which does not require the installation of dedicated wiring for the transmission of its alarm signal.

Positive Noninterfering (PNI) and Successive Alarm System—An alarm system which employs multiple alarm transmitters on each alarm line (like a McCulloch loop) such that in the event of simultaneous operation of several transmitters, one of them takes control of the alarm line, transmits its full signal, then releases the alarm line for successive transmission by other transmitters which are held inoperative until they gain control.

Pressure Alarm System—An alarm system which protects a vault or other enclosed space by maintaining and monitoring a predetermined air pressure differential between the inside and outside of the space. Equalization of pressure resulting from opening the vault or cutting through the enclosure will be sensed and will initiate an alarm signal.

Printing Recorder—An electromechanical device used at a monitoring station which accepts coded signals from alarm lines and converts them to an alphanumeric printed record of the signal received.

Proprietary Alarm System—An alarm system which is similar to a central station alarm system except that the annunciator is located in a constantly manned guard room maintained by the owner for his own internal security operations. The guards monitor the system and respond to all alarm signals or alert local law enforcement agencies or both.

Protected Area—An area monitored by an alarm system or guards, or enclosed by a suitable barrier.

Protected Port—A point of entry such as a door, window, or corridor which is monitored by sensors connected to an alarm system.

Protection Device—(1) A sensor such as a grid, foil, contact, or photoelectric sensor connected into an intrusion alarm system. (2) A barrier which inhibits intrusion, such as a grille, lock, fence or wall.

Protection, Exterior Perimeter—A line of protection surrounding but somewhat removed from a facility. Examples are fences, barrier walls, or patrolled points of a perimeter.

Protection Off—*See* Access Mode.

Protection On—*See* Secure Mode.

Protective Screen—*See* Grid.

Protective Signaling—The initiation, transmission, and reception of signals involved in the detection and prevention of property loss due to fire, burglary, or other destructive conditions. Also, the electronic supervision of persons and equipment concerned with this detection and prevention. *See also* Line Supervision and Supervisory Alarm System.

Proximity Alarm System—*See* Capacitance Alarm System.

Punching Register—*See* Register, Punch.

Radar Alarm System—An alarm system which employs radio frequency motion detectors.

Radar (Radio Detecting and Ranging)—*See* Radio Frequency Motion Detector.

Radio Frequency Interference (RFI)—Electromagnetic interference in the radio frequency range.

Radio Frequency Motion Detector—A sensor which detects the motion of an intruder through the use of a radiated radio frequency electromagnetic field. The device operates by sensing a disturbance in the generated rf field caused by intruder motion. Typically a modulation of the field is referred to as a doppler effect, and is used to initiate an alarm signal. Most radio frequency motion detectors are certified by the FCC for operation as "field disturbance sensors" at one of the following frequencies: 0.915 GHz (L-Band), 2.45 GHz (S-Band), 5.8 GHz (X-Band), 10.525 GHz (X-Band), and 22.125 GHz (K-Band). Units operating in the microwave frequency range are usually called microwave motion detectors.

Reed Switch—A type of magnetic switch consisting of contacts formed by two thin movable magnetically actuated metal vanes or reeds, held in a normally open position within a sealed glass envelope.

Register—An electromechanical device which marks a paper tape in response to signal impulses received from transmitting circuits. A register may be driven by a prewound spring mechanism, an electric motor, or a combination of these.

Register, Inking—A register which marks the tape with ink.

Register, Punch—A register which marks the tape by cutting holes in it.

Register, Slashing—A register which marks the tape by cutting V-shaped slashes in it.

Remote Alarm—An alarm signal which is transmitted to a remote monitoring station. *See also* Local Alarm.

Remote Station Alarm System—An alarm system which employs remote alarm stations usually located in building hallways or on city streets.

Reporting Line—*See* Alarm Line.

Reset—To restore a device to its original (normal) condition after an alarm or trouble signal.

Resistance Bridge Smoke Detector—A smoke detector which responds to the particles and moisture present in smoke. These substances reduce the resistance of an electrical bridge grid and cause the detector to respond.

Retard Transmitter—A coded transmitter in which a delay period is introduced between the time of actuation and the time of signal transmission.

RFI—Radio Frequency Interference.

Rf Motion Detector—*See* Radio Frequency Motion Detector.

Robbery—The felonious or forcible taking of property by violence, threat, or other overt felonious act in the presence of the victim.

Secure Mode—The condition of an alarm system in which all sensors and control units are ready to respond to an intrusion.

Security Monitor—*See* Annunciator.

Seismic Sensor—A sensor, generally buried under the surface of the ground for perimeter protection, which responds to minute vibrations of the earth generated as an intruder walks or drives within its detection range.

Sensor—A device which is designed to produce a signal or offer indication in response to an event or stimulus within its detection zone.

Sensor, Combustion—*See* Ionization Smoke Detector, Photoelectric Beam-Type Smoke Detector, Photoelectric Spot-Type Smoke Detector, and Resistance Bridge Smoke Detector.

Sensor, Smoke—*See* Ionization Smoke Detector, Photoelectric Beam-Type Smoke Detector, Photoelectric Spot-Type Smoke Detector, and Resistance Bridge Smoke Detector.

Shunt—(1) A deliberate shorting-out of a portion of an electric circuit. (2) A key-operated switch which removes some portion of an alarm system for operation, allowing entry into a protected area without initiating an alarm signal. A type of authorized access switch.

Shunt Switch—*See* Shunt.

Signal Recorder—*See* Register.

Silent Alarm—A remote alarm without an obvious local indication that an alarm has been transmitted.

Silent Alarm System—An alarm system which signals a remote station by means of a silent alarm.

Single Circuit System—An alarm circuit which routes only one side of the circuit through each sensor. The return may be through either ground or a separate wire.

Single-Stroke Bell—A bell which is struck once each time its mechanism is activated.

Slashing Register—*See* Register, Slashing.

Smoke Detector—A device which detects visible or invisible products of combustion. *See also* Ionization Smoke Detector, Photoelectric Beam-Type Smoke Detector, Photoelectric Spot-Type Smoke Detector, and Resistance Bridge Smoke Detector.

Solid State—(1) An adjective used to describe a device such as a semiconductor transistor or diode. (2) A circuit or system which does not rely on vacuum or gas-filled tubes to control or modify voltages and currents.

Sonic Motion Detector—A sensor which detects the motion of an intruder by his disturbance of an audible sound pattern generated within the protected area.

Sound Sensing Detector System—An alarm system which detects the audible sound caused by an attempted forcible entry into a protected structure. The system consists of microphones and a control unit containing an amplifier, accumulator, and a power supply. The unit's sensitivity is adjustable so that ambient noises or normal sounds will not initiate an alarm signal.

However, noises above this preset level or a sufficient accumulation of impulses will initiate an alarm.

Sound Sensor—A sensor which responds to sound; a microphone.

Space Protection—*See* Area Protection.

Spoofing—The defeat or compromise of an alarm system by "tricking" or "fooling" its detection devices, such as by short-circuiting part or all of a series circuit, cutting wires in a parallel circuit, reducing the sensitivity of a sensor, or entering false signals into the system. Spoofing contrasts with circumvention.

Spot Protection—Protection of objects such as safes, art objects, or anything of value which could be damaged or removed from the premises.

Spring Contact—A device employing a current-carrying cantilever spring which monitors the position of a door or window.

Standby Power Supply—Equipment which supplies power to a system in the event the primary power is lost. It may consist of batteries, charging circuits, auxiliary motor generators or a combination of these devices.

Strain Gauge Alarm System—An alarm system which detects the stress caused by the weight of an intruder as he moves about a building. Typical uses include placement of the strain gauge sensor under a floor joist or under a stairway tread.

Strain Gauge Sensor—A sensor which, when attached to an object, will provide an electrical response to an applied stress upon the object, such as a bending, stretching or compressive force.

Strain Sensitive Cable—An electrical cable which is designed to produce a signal whenever the cable is strained by a change in applied force. Typical uses include mounting it in a wall to detect an attempted forced entry through the wall, or fastening it to a fence to detect climbing on the fence, or burying it around a perimeter to detect walking or driving across the perimeter.

Subscriber's Equipment—That portion of a central station alarm system installed in the protected premises.

Subscriber's Unit—A control unit of a central station alarm system.

Supervised Lines—Interconnecting lines in an alarm system which are electrically supervised against tampering. *See also* Line Supervision.

Supervisory Alarm System—An alarm system which monitors conditions or persons or both, and signals any deviation from an established norm or schedule. Examples are the monitoring of signals from guard patrol stations for irregularities in the progression along a prescribed patrol route, and the monitoring of production or safety conditions such as sprinkler water pressure, temperature, or liquid level.

Supervisory Circuit—An electrical circuit or radio path which sends information on the status of a sensor or guard patrol to an annunciator. For intrusion alarm systems, this circuit provides line supervision and monitors tamper devices. *See also* Supervisory Alarm System.

Surreptitious—Covert, hidden, concealed, or disguised.

Surveillance—(1) Control of premises for security purposes through alarm systems, closed circuit television (cctv), or other monitoring methods. (2) Supervision or inspection of industrial processes by monitoring those conditions which could cause damage if not corrected. *See also* Supervisory Alarm System.

Tamper Device—(1) Any device, usually a switch, which is used to detect an attempt to gain access to intrusion alarm circuitry, such as by removing a switch cover. (2) A monitor circuit to detect any attempt to modify the alarm circuitry, such as by cutting a wire.

Tamper Switch—A switch which is installed in such a way as to detect attempts to remove the enclosure of some alarm system components such as control box doors, switch covers, junction box covers, or bell housings. The alarm component is then often described as being "tampered."

Tape—*See* Foil.

Tapper Bell—A single-stroke bell designed to produce a sound of low intensity and relatively high pitch.

Telephone Dialer, Automatic—A device which, when activated, automatically dials one or more pre-programmed telephone numbers (e.g., police, fire department) and relays a recorded voice or coded message giving the location and the nature of the alarm.

Telephone Dialer, Digital—An automatic telephone dialer which sends its message as a digital code.

Terminal Resistor—A resistor used as a terminating device.

Terminating Capacitor—A capacitor sometimes used as a terminating device for a capacitance sensor antenna. The capacitor allows the supervision of the sensor antenna, especially if a long wire is used as the sensor.

Terminating Device—A device which is used to terminate an electrically supervised circuit. It makes the electrical circuit continuous and provides a fixed impedance reference (end of line resistor) against which changes are measured to detect an alarm condition. The impedance changes may be caused by a sensor, tampering, or circuit trouble.

Time Delay—*See* Entrance Delay and Exit Delay.

Time Division Multiplexing (TDM)—*See* Multiplexing, Time Division.

Timing Table—That portion of central station equipment which provides a means for checking incoming signals from McCulloh Circuits.

Touch Sensitivity—The sensitivity of a capacitance sensor at which the alarm device will be activated only if an intruder touches or comes in very close proximity (about 1 cm or ½ in) to the protected object.

Trap—(1) A device, usually a switch, installed within a protected area, which serves as secondary protection in the event a perimeter alarm system is successfully penetrated. Examples are a trip wire switch placed across a likely path for an intruder, a match switch hidden under a rug, or a magnetic switch mounted on an inner door. (2) A volumetric sensor installed so as to detect an intruder in a likely traveled corridor or pathway within a security area.

Trickle Charge—A continuous direct current, usually very low, which is applied to a battery to maintain it at peak charge or to recharge it after it has been partially or completely discharged. Usually applied to nickel cadmium (Nicad) or wet cell batteries.

Trip Wire Switch—A switch which is actuated by breaking or moving a wire or cord installed across a floor space.

Trouble Signal—*See* Break Alarm.

UL—*See* Underwriters Laboratories, Inc.

UL Certificated—For certain types of products which have met UL requirements, for which it is impractical to apply the UL Listing Mark or Classification Marking to the individual product, a certificate is provided which the manufacturer may use to identify quantities of material for specific job sites or to identify field installed systems.

UL Listed—Signifies that production samples of the product have been found to comply with established Underwriters Laboratories requirements and that the manufacturer is authorized to use the Laboratories' Listing Marks on the listed products that comply with the requirements, contingent upon the follow-up services as a check of compliance.

Ultrasonic—Pertaining to a sound wave having a frequency above that of audible sound (approximately 20,000 Hz). Ultrasonic sound is used in ultrasonic detection systems.

Ultrasonic Detection System—See Ultrasonic Motion Detector and Passive Ultrasonic Alarm System.

Ultrasonic Frequency—Sound frequencies which are above the range of human hearing; approximately 20,000 Hz and higher.

Ultrasonic Motion Detector—A sensor which detects the motion of an intruder through the use of ultrasonic generating and receiving equipment. The device operates by filling a space with a pattern of ultrasonic waves; the modulation of these waves by a moving object is detected and initiates an alarm signal.

Underdome Bell—A bell most of whose mechanism is concealed by its gong.

Underwriters Laboratories, Inc. (UL)—A private independent research and testing laboratory which tests and lists various items meeting good practice and safety standards.

Vibrating Bell—A bell whose mechanism is designed to strike repeatedly and for as long as it is activated.

Vibrating Contact—See Vibration Sensor.

Vibration Detection System—An alarm system which employs one or more contact microphones or vibration sensors which are fastened to the surfaces of the area or object being protected to detect excessive levels of vibration. The contact microphone system consists of microphones, a control unit containing an amplifier and an accumulator, and a power supply. The sensitivity of the unit is adjustable so that ambient noises or normal vibrations will not initiate an alarm signal. In the vibration sensor system, the sensor responds to excessive vibration by opening a switch in a closed circuit system.

Vibration Detector—See Vibration Sensor.

Vibration Sensor—A sensor which responds to vibrations of the surface on which it is mounted. It has a normally closed switch which will momentarily open when it is subjected to a vibration with sufficiently large amplitude. Its sensitivity is adjustable to allow for the different levels of normal vibration, to which the sensor should not respond, at different locations. *See also* Vibration Detection System.

Visual Signal Device—A pilot light, annunciator or other device which provides a visual indication of the condition of the circuit or system being supervised.

Volumetric Detector—See Volumetric Sensor.

Volumetric Sensor—A sensor with a detection zone which extends over a volume such as an entire room, part of a room, or a passageway. Ultrasonic motion detectors and sonic motion detectors are examples of volumetric sensors.

Walk Test Light—A light on motion detectors which comes on when the detector senses motion in the area. It is used while setting the sensitivity of the detector and during routine checking and maintenance.

Watchman's Reporting System—A supervisory alarm system arranged for the transmission of a patrolling watchman's regularly recurrent report signals from stations along his patrol route to a central supervisory agency.

Zoned Circuit—A circuit which provides continual protection for parts or zones of the protected area while normally used doors and windows or zones may be released for access.

Zones—Smaller subdivisions into which large areas are divided to permit selective access to some zones while maintaining other zones secure and to permit pinpointing the specific location from which an alarm signal is transmitted.

Index

O

P

U

V

W